Vapor Compression Heat Pumps

with Refrigerant Mixtures

Vapor Compression Heat Pumps

with Refrigerant Mixtures

Reinhard Radermacher
Yunho Hwang

CRC Press
Taylor & Francis Group
Boca Raton London New York

CRC Press is an imprint of the
Taylor & Francis Group, an **informa** business
A TAYLOR & FRANCIS BOOK

Published in 2005 by
CRC Press
Taylor & Francis Group
6000 Broken Sound Parkway NW, Suite 300
Boca Raton, FL 33487-2742

First issued in paperback 2019

No claim to original U.S. Government works

ISBN-13: 978-0-367-45411-1 (pbk)
ISBN-13: 978-0-8493-3489-4 (hbk)

Library of Congress Card Number 2005042099

Library of Congress Cataloging-in-Publication Data

Radermacher, Reinhard.
   Vapor compression heat pumps with refrigerant mixes / Reinhard Radermacher, Yunho Hwang.
         p. cm.
   Includes bibliographical references and index.
   ISBN 0-8493-3489-6 (alk. paper)
   1. Heat pumps. 2. Refrigerants. I. Hwang, Yunho. II. Title.

TJ262.R33 2005
621.5′63--dc22

                                                                          2005042099

Visit the Taylor & Francis Web site at
http://www.taylorandfrancis.com

and the CRC Press Web site at
http://www.crcpress.com

Preface

The authors strive to provide a comprehensive background and thorough understanding of the thermodynamics of working fluid mixtures and their applications in vapor compression systems. The information presented here is based on an extensive review of the literature and includes a considerable body of experience gained by the authors from over two decades of research in the laboratories of CEEE at the University of Maryland, the National Institute of Standards and Technology (NIST), and the Oak Ridge National Laboratory (ORNL). To ensure that this text is comprehensive and up to date, the authors invite all interested parties to submit corrections, recommendations, and suggestions for improvement to Reinhard Radermacher at raderm@eng.umd.edu or Yunho Hwang at yhhwang@eng.umd.edu.

Acknowledgment

The partial support of this effort by the U.S. Department of Energy through Oak Ridge National Laboratory (ORNL) and the University of Maryland is gratefully acknowledged.

About the Authors

Dr. Yunho Hwang, a research associate professor at the University of Maryland's department of mechanical engineering, works with Dr. Radermacher in the Center for Environmental Energy Engineering. Dr. Hwang's career began as a senior researcher at Samsung Electronics in Korea, and he has devoted more than 20 years to university and private industry research. Dr. Yunho Hwang, Director of the Heat Pump Laboratory (CEEE), is an internationally recognized expert in energy conversion systems; in particular, heat pumping, air conditioning, and refrigeration systems. He has coauthored 3 books and written more than 50 academic papers. He currently serves as secretary of the IIR (International Institute of Refrigeration) Commission B2.

Dr. Reinhard Radermacher is a professor of mechanical engineering at the University of Maryland and director of the university's Center for Environmental Energy Engineering (CEEE). His engineering expertise, spanning over two decades, is concentrated in energy conversion, renewable energy, heating and power systems, and air conditioning and refrigeration.

Dr. Radermacher is a graduate of the Munich Institute of Technology, Germany, and has worked with numerous private energy corporations, as well as government institutions and engineering groups. He currently serves as editor of ASHRAE's (American Society of Heating, Refrigerating and Air-Conditioning Engineers) *HVAC&R Research Journal* and associate editor of the *International Journal of Refrigeration*. The author of more than 100 academic papers, Dr. Radermacher is also coauthor of the books *Heat Conversion Systems* and *Absorption Heat Pumps and Chillers*, both published by CRC Press.

NOMENCLATURE

17L	Low Temperature Heating Test Based on ASHRAE Standard 116
35F	Frost Accumulation Test Based on ASHRAE Standard 116
47S	High Temperature Heating Test Based on ASHRAE Standard 116
47C	Cyclic Heating Test Based on ASHRAE Standard 116
A	Heat Exchanger Area
AREP	Alternative Refrigerant Evaluation Program
ARI	Air-Conditioning and Refrigeration Institute, Arlington, VA
ASHRAE	American Society of Heating, Refrigerating and Air-Conditioning Engineers, Inc.
AT	Approach Temperature
Bo	Boiling Number
C_D	Cyclic Degradation Coefficient
CFC	Chlorofluorocarbon
CLF	Cooling Load Factor
CP	Critical Point
c_p	Isobaric Specific Heat
COP	Coefficient of Performance
CSPF	Cooling Seasonal Performance Factor
D	Tube Diameter
DOE	Department of Energy
DGWP	Direct Global Warming Potential
E	Enhancement Factor
EB	Energy Balance between Water-side and Refrigerant-side
EPA	Environmental Protection Agency
f	Frictional Coefficient
Fr	Froude Number
G	Mass Flux
g	Acceleration of Gravity
GWP	Global Warming Potential
h	Enthalpy or Heat Transfer Coefficient
HCFC	Hydrochlorofluorocarbon
HFC	Hydrofluorocarbon
HLF	Heating Load Factor
HSPF	Heating Seasonal Performance Factor
HTC	Heat Transfer Coefficient
HX	Heat Exchanger
ID	Inside Diameter
IDLH	Immediately Dangerous to Life or Health, maximum level from which one could escape within 30 minutes without impairing symptoms or any irreversible health effects
IGWP	Indirect Global Warming Potential

k	Thermal Conductivity
LMTD	Log Mean Temperature Difference
LPE	Liquid Precooling in Evaporator
M	Molecular Weight
$\dot{m}$	Mass Flow Rate
N	Number or Polytropic Index
NIST	National Institute of Science and Technology
NTU	Number of Transfer Unit
Nu	Nusselt Number
OD	Outside Diameter
ODP	Ozone Depletion Potential
ORNL	Oak Ridge National Laboratory
P	Pressure
P_r	Reduced Pressure
Pr	Prandtl Number
Q	Capacity
R407C	R32/R125/R134a mixture in compositions of 23/25/52 wt.%
R410A	R32/R125 (50/50 wt.%)
R410B	R32/R125 (45/55 wt.%)
Re	Reynolds Number
RT	Refrigeration Ton
s	Entropy
S	Suppression Factor
SEER	Seasonal Energy Efficiency Ratio
SPF	Seasonal Performance Factors
SLHX	Suction-Line Heat Exchanger
ST	Short Tube Restrictor
T	Temperature
TEV	Thermostatic Expansion Valve
TLV	Threshold Limit Value, the refrigeration concentration limit in the air for a normal 8-hour workday that will not cause an adverse effect in most people.
U	Overall Heat Transfer Coefficient
VFR	Volume Flow Rate
X_{tt}	Martinelli Parameter
x	Mass Fraction
x_q	Vapor Quality
ΔP	Pressure Drop

SUBSCRIPT DESIGNATIONS

a   air
amb   ambient
avg   average
b   bulk flow or bubble point
bc   bulk convection
c   cross-section
cp   critical point
cal   calculated
co   condenser
cyc   cyclic
d   dew point
dis   compressor discharge
e   evaporation
ev   evaporator
exp   experimental
f   fin
fg   liquid vapor
gc   gas cooler
i   mean integrated
in   inlet
isen   isentropic
l   liquid
mec   mechanical
MN   Martinelli-Nelson
mot   motor
nb   nucleate boiling
o   wall or overall
opt   optimum
out   outlet
pol   polytropic
pool   pool boiling
r   refrigerant
sat   saturation
sub   subcooling at the condenser outlet
suc   compressor suction
sup   superheating at the evaporator outlet
sv   saturated vapor
t   total
tp   two-phase
v   vapor
vol   volumetric
w   water or wall

GREEK SYMBOLS

μ dynamic viscosity

ρ density

σ surface tension

ϕ two-phase heat transfer multiplier

α void fraction, $A_{vapor} / A_{cross\text{-}sectional}$

Table of Contents

1 Introduction

Due to growing environmental awareness and resulting concerns, refrigerants — the working fluids for refrigeration systems, heat pumps and air conditioners — have gained considerable attention. Well-known, proven fluids were banned and replaced with new ones. In the process, refrigerant mixtures were introduced to achieve acceptable properties reasonably well-matched to those of the fluids to be replaced. The thermodynamics of mixtures is considerably more complex than that of pure fluids. This text is dedicated to introducing the thermodynamic theory and background of working fluid mixtures and their practical implementations. It discusses the opportunities and challenges and aims at providing the practicing engineer, as well as the aspiring engineering student, with all the necessary information to successfully design systems that take the best possible advantage of fluid mixtures. Emphasis is placed on practical experience from laboratory investigations.

1.1 HEAT PUMPING

On many occasions, a fluid needs to be cooled below the temperature of the surroundings. Examples are the production of chilled water or brine for air-conditioning and refrigeration purposes or the cooling of air to condition a space. In this context, the term "cooling" usually implies a certain quantity of energy is extracted from a fluid while reducing its temperature. This energy is rejected to a second fluid at a higher temperature than that of the first fluid. For example, in a domestic refrigerator, energy is extracted from the air inside the cabinet, reducing its temperature and cooling the cabinet. In turn, energy is rejected from the refrigeration unit to the surrounding air, heating the exterior. When energy transfer is based on a temperature gradient, one speaks of heat transfer. In this strict sense, "heat" exists only as long as a temperature gradient enables this transfer of energy. However, in the field of refrigeration and air-conditioning, the term "heat" is used more loosely and is understood as energy in more general terms. This text adheres to this latter convention.

In general, the machines that accomplish the lift of a certain quantity of energy from a lower temperature level to a higher one are termed *heat pumps*. The thermodynamic processes involved in heat pumping are governed by the First and Second Laws of Thermodynamics. The First Law is the conservation of energy and reads

$$\sum Q_i + W = 0 \qquad (1.1)$$

where Q stands for the amounts of heat exchanged at various temperature levels T_i and W for the amount of work that is required in the process.

The Second Law states that heat cannot flow from a lower to a higher temperature without the expenditure of energy. One way of expressing this statement is shown in the following equation.

$$\sum \frac{Q_i}{T_i} = 0 \quad \text{for reversible processes} \qquad (1.2)$$

Equation 1.1 is based on the assumption that the energy supplied to a system is counted as positive.

Figure 1.1.1 illustrates the situation. The vertical axis is the temperature axis with increasing temperature from the bottom to top, i.e., $T_1 < T_2$. A certain amount of heat is removed by the device M at temperature T_1 and, by using work that is supplied to M, heat is rejected at T_2. The amount of heat rejected is the sum of the amount of work W required by M and the amount of heat removed at T_1. This is a consequence of the First Law,

$$Q_1 + W = Q_2 \qquad (1.3)$$

However, does this relationship of Figure 1.1.1 fulfill the Second Law, Equation 1.2? Applying Equation 1.2 to the situation of Figure 1.1.1 yields

$$\frac{Q_1}{T_1} - \frac{Q_2}{T_2} = 0 \qquad (1.4)$$

and using Equation 1.3,

$$\frac{Q_1}{T_1} - \frac{Q_1 + W}{T_2} = 0 \qquad (1.5)$$

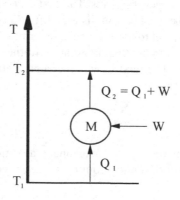

FIGURE 1.1.1 Heat Pumping — The process of lifting energy Q_1 from a temperature level T_1 to T_2, releasing Q_2. The input of energy, in this case work W, is required.

After rearranging the terms, we find

$$\frac{Q_1}{T_1} - \frac{Q_1}{T_2} - \frac{W}{T_2} = 0 \qquad (1.6)$$

Since $T_2 > T_1$, the difference of the first two terms is still larger than zero. Equation 1.6 requires that the Second Law is fulfilled only if the work input is sufficiently large. Thus, there is a minimum work requirement. This minimum amount of work is required for the system to operate reversibly. The Second Law does not say anything about an upper limit for the work, so the system may be very inefficient and operate irreversibly if W becomes much larger than necessary. There is no upper limit on the degree of irreversibility for the heat pumping operation.

The First and Second Laws state for lifting heat from one temperature level to a higher one, the expenditure of energy is required and that the amount of energy rejected at the higher temperature level is the sum of the heat removed from the lower temperature level plus the energy added to accomplish the lift. The energy required to drive the system can be supplied in the form of work, as assumed here, or it can be supplied in the form of heat. Examples are absorption heat pumps, desiccant systems and engine-driven heat pumps.

One of the most common implementations of a heat pump is shown in Figure 1.1.2. Liquid refrigerant evaporates in the evaporator at a given pressure level while it absorbs heat. In other terms, this is the *cooling effect*. The compressor compresses the vapor that exits from the evaporator and delivers the compressed, high-pressure vapor to the condenser where the vapor is cooled and condensed. The liquefied vapor leaves the condenser and enters the expansion valve. This valve reduces the pressure level and meters the fluid so that as much refrigerant enters the evaporator as the compressor removes.

The concept shown in Figure 1.1.2 represents a vapor compression heat pump. It is the underlying concept of the type of heat pumps discussed in this text. All of these systems have to fulfill the First and Second Law in the form of the Equations 1.1 and 1.2.

It should be noted that although both the First and Second Law must be observed in heat pump design, the reader will notice that in subsequent chapters energy and

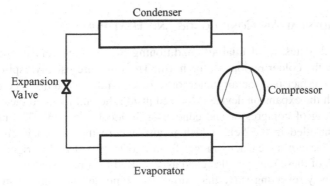

FIGURE 1.1.2 Schematic of a vapor compression heat pump (or refrigeration system).

mass balances are discussed but rarely entropy balances. This implies that the Second Law could be violated in idealized situations. However, in all examples we are using measured data of actual working fluids that do exist. These data inherently ensure that the Second Law is not violated.

The following sections provide an overview of various heat pump applications. Note that whenever energy is lifted from a lower temperature to a higher temperature, we speak in general of a heat pumping process. However, for many applications, the process is not as important as its result. In all refrigeration and air-conditioning applications, energy is removed from a lower temperature level and rejected at a higher temperature level. The main effect that interests the user is the refrigeration or cooling effect, i.e., heat removal. On the other hand, in what is customarily called heat pumping, the user is interested in the heat delivered at the high temperature level. In both cases, the underlying thermodynamic process is the same; only the application or the desired benefit are different. In this text, the term *heat pumping* is used as the general description for the thermodynamic process independent of the application. The terms *refrigeration* and *air-conditioning* refer specifically to the application.

1.2   OVERVIEW OF CURRENT PRODUCTS

Vapor compression heat pump systems have found many applications spanning a wide range of capacities. This chapter will give an overview of applications, system configurations and the economic importance of vapor compression systems.

The global market for air-conditioning and refrigeration equipment is estimated to be $42 to 45 billion, with the United States and Japan each covering one third of the worldwide production followed by Europe and China. In terms of applications, the market is split into thirds as well. One third covers residential air-conditioning, one-third refrigeration (mostly food preservation), and one-third commercial air-conditioning. In the United States, stationary refrigeration and air-conditioning systems are used for 45 million homes and commercial buildings as well as 100 million refrigerators and 30 million freezers. Mobile refrigeration and air-conditioning systems are used for 90 million air-conditioned cars and trucks and 200,000 refrigerated trucks and rail cars. In 1985, approximately 420 million pounds of refrigerants were used for these applications.

1.2.1   RESIDENTIAL AIR CONDITIONERS AND HEAT PUMPS

In the United States, residential air-conditioning has become very common and is even used in the colder regions of the north. The units are usually "split" systems, referring to the fact that the air conditioner is split into two units. The evaporator together with the expansion device is located inside the building and the condensing unit, consisting of compressor and condenser, is located outside. The piping connection is installed in the field, which is also where the system is charged with refrigerant. The cooling capacity ranges from 3 to 18 kW (1 to 5 refrigeration tons). Around 18% of these units are designed as so-called heat pumps (ARI, 2000). They have a four-way reversing valve that allows the same unit to serve as an air conditioner in the summer months and as a heat pump in the winter. Since traditionally the system is designed and sized for the air-conditioning function, the heat pumps turn out to have insufficient heating capacity in most climates. Supplemental heat in the form of electric or gas heat is usually required.

Another form of residential air-conditioning is the window air conditioner. It is designed to cool just one room and the largest capacity usually does not exceed 7 kW (2 refrigeration tons). All the components are integrated within one chassis that is usually mounted within the frame of a window.

1.2.2 COMMERCIAL AIR-CONDITIONING SYSTEMS

Commercial air-conditioning systems span a wide range of capacities from 17 kW (5 refrigeration tons) to several thousand tons. The larger ones usually employ variable speed centrifugal compressors and low-pressure refrigerants for high efficiencies. They are designed to produce chilled water at about 5 to 7°C and are cooled by water that in turn rejects heat to the ambient in a wet cooling tower. The chilled water is supplied to air-handling units located throughout the building.

It is interesting to note that many commercial buildings require cooling year-round because these buildings have considerable internal heat loads resulting from the presence of occupants and equipment that generate heat. A considerable floor area of useful space is located within the core of the building. This interior space is surrounded by other rooms that have outside walls but these perimeter rooms are always maintained at a constant temperature. Thus, the interior space has no way of rejecting heat and requires cooling year-round. In newer systems, a so-called economizer allows the use of outdoor air for core cooling whenever weather conditions are suitable.

Recently, a new generation of chillers emerged that are vapor compression cycles driven by internal combustion engines. The primary energy input is gas rather than electricity and the new generation chillers reduce the reliance on costly peak power and help to avoid high demand charges.

For smaller buildings or those with a limited number of stories, so-called rooftop units are used. These are air-conditioning systems that reach up to 700 kW (200 refrigeration tons) in capacity and have the air handler integrated within the unit, often including a heating device such as a gas furnace.

1.2.3 INDUSTRIAL SYSTEMS

Industrial refrigeration systems serve a wide range of purposes. They include food storage in large warehouses to the provision of cold utilities in chemical plants, refineries and polymer production facilities. Further, many production facilities such as printing, spinning, electronics and precision machining require very stringently controlled thermal environments. Many machines for industrial refrigeration and air-conditioning are custom-manufactured to meet specific criteria.

1.2.4 REFRIGERATION SYSTEMS

Probably the most important and earliest application of refrigeration technology was that of food storage. Today, the technology has developed into a complete cold chain in which food products such as fish, meats, dairy products and vegetables are cooled or refrigerated from the moment of harvest or production through storage, transport, wholesale and retail to the consumer, who uses the residential refrigerator as the last storage facility in this chain. Accordingly, a wide range of refrigeration systems have been developed that meet the various needs.

The domestic refrigerator is a unique device in that it has reached such a level of maturity that some units last more than 20 years without ever failing or requiring maintenance. In addition, the energy consumption has been reduced by more than a factor of two over the past 30 years. The newest models of a typical refrigerator for a U.S. household show a power consumption that is equivalent to that of a 50 W light bulb left on continuously.

Refrigeration systems in supermarkets usually have central compressor plants with multiple compressors in parallel that serve the same set of evaporators. The evaporators are located throughout the store in the display cabinets for frozen or refrigerated foods. As a consequence, a single store can boast miles of refrigerant piping. Due to the length of piping and the many inherent piping connections, supermarket refrigeration systems tended to lose considerable amounts of refrigerant. With the advent of environmental concerns and improved quality standards, these leakages have been reduced considerably. New systems are under consideration in which the refrigerant does not leave the machine room. Rather, a secondary refrigerant is used to cool the display cabinets. This measure reduces refrigerant leakage and increases system reliability, but may contribute to a reduction in energy efficiency.

Transport refrigeration systems, especially those that are trailer or container mounted, must come with their own power supply, usually a small diesel or gas engine. Systems must be especially robust to withstand the high vibration levels. Also, these systems must be designed to cover a wide range of operating conditions. The transported goods may require storage temperatures ranging from −40 to 20°C while the ambient temperature can span even a wider range of temperatures.

For food preservation, there are also a wide array of specialized required cooling systems. Examples are milk coolers on farms, cooling systems for tank trucks and refrigeration systems ranging from fish trawlers to wine coolers.

The air-conditioning of cars has become an expected feature in the United States and many developed and developing countries. The cooling capacity varies over a wide range because the compressor is directly coupled to the engine and the compressor revolutions vary with that of the engine. Furthermore, due to vibration and the need to use flexible hoses to connect the components that are mounted on the car and those mounted on the engine block, these systems traditionally have a relatively high rate of refrigerant leakage. With the advent of environmental concerns, the leakage rates have been reduced dramatically.

1.3 HISTORY OF WORKING FLUIDS

The First and Second Laws state only the conditions and limitations under which machinery such as refrigeration systems can operate. The actual implementation is not prescribed by these laws. It is common practice in the refrigeration industry to employ the vapor compression cycle of Figure 1.1.2. To do so requires the use of working fluids, commonly termed *refrigerants*. They represent one of the key ingredients of any vapor compression system and influence design, operation and cost immensely. A short overview of the historical development of refrigerants follows that provides insight into the development of the refrigeration systems. The following is an excerpt of a paper presented by Calm and Didion in 1997.

Refrigeration goes back to ancient times, using stored ice and a number of evaporative processes. Early in the 19th century, the use of a volatile fluid in a closed cycle was proposed by Oliver Evans. He used evaporating ether, under a vacuum, and then pumped the vapor to a water-cooled heat exchanger to be condensed for reuse.

Other refrigerants were introduced in the 1830s with the invention of the vapor-compression machine by Perkins. He designed a machine to use sulfuric (ethyl) ether as the refrigerant (Perkins, 1834). The first machine actually used caoutchoucine, an industrial solvent that Perkins apparently utilized in his business as a printer. It seems the first tradeoff in refrigerants — and one still driving selections — was based on availability.

Table 1.1 summarizes the historic introduction of refrigerants. The years given serve only as guidelines, since the period from the introduction of a new fluid to its generally accepted use always spans a number of years. Table 1.1 summarizes early refrigerants, namely those predating fluorinated chemicals and state-of-the-art fluoro-chemicals that are under consideration today. Downing (1988), Nagengast (1989 and 1996) and Thévenot (1979) present further details.

The first century of refrigerant use was dominated by innovative efforts with familiar fluids. The goals were to provide refrigeration and, later, to improve machine durability. Use of blends was attempted where single-compound solutions could not be found (Pictet, 1885). After World War I, attention turned to safety and performance as well. Willis H. Carrier and R.W. Waterfill initiated one of the first documented systematic searches (Carrier and Waterfill, 1924). They investigated a range of candidate refrigerants for suitability for both positive displacement and centrifugal compression machines. These analyses closely examined ammonia, ethyl ether, carbon dioxide, carbon tetrachloride, sulfur dioxide and water. They finally selected dielene (1,2-dichloroethene, R1130) for the first centrifugal machine, though an international search was needed to find a source for it (Ingels, 1952).

Nearly all of the early refrigerants were flammable, toxic or both, and some were also highly reactive. Accidents were common. For perspective, propane was marketed as the odorless safety refrigerant (CLPC, 1922).

The discovery of fluorinated refrigerants began with Thomas Midgley, Jr., in 1928 (Midgley, 1937). Midgley and his associates, Albert L. Henne and Robert R. McNary, scoured property tables to find chemicals with the desired boiling point. They restricted the search to those known to be stable but not toxic or flammable. They eventually used organic fluorides. While fluorine was known to be toxic by itself, Midgley and his collaborators felt that compounds containing it could be nontoxic.

Recognizing the deficiencies of the published literature, Midgley, Henne and McNary turned to the periodic table of the elements. They quickly eliminated those elements yielding insufficient volatility and then eliminated those resulting in unstable

TABLE 1.1
Introduction of Refrigerants

Year	Refrigerant	Chemical Makeup, Formula
1830s	Caoutchoucine	Distillate of india rubber
		$CH_3-CH_2-O-CH_2-CH_3$
1840s	Methyl ether (RE170)	CH_3-O-CH_3
1850	Water/sulfuric acid	H_2O/H_2SO_4
1856	Ethyl alcohol	CH_3-CH_2-OH
1859	Ammonia/water	NH_3/H_2O
1866	Chymogene	Petrol ether and naphtha (hydrocarbons)
	Carbon dioxide	CO_2
1860s	Ammonia (R717)	$NH_3CH_3(NH_2)$
	Methyl amine (R630)	$CH_3-CH_2(NH_2)$
	Ethyl amine (R631)	
1870	Methyl formate (R611)	$HCOOCH_3$
1875	Sulfur dioxide (R764)	SO_2
1878	Methyl chloride (R40)	CH_3CL
1870s	Ethyl chloride (R160)	CH_3-CH_2CL
1891	Blends of sulfuric acid with hydrocarbons	H_2SO_4, C_4H_{10}, C_5H_{12}, $(CH_3)_2CH-CH_3$
1900s	Ethyl bromide (R160B1)	CH_3-CH_2Br
1912	Carbon tetrachloride	CCl_4
	Water vapor (R718)	H_2O
1920s	Isobutane (R600a)	$(CH_3)_2CH-CH_3$
	Propane (R290)	$CH_3-CH_2-CH_3$
1922	Dielene (R1130^a)	$CHCl=CHCl$
1923	Gasoline	Hydrocarbons
1925	Trielene (R1120)	$CHCl=CCl_2$
1926	Methylene chloride (R30)	CH_2Cl_2
1931	R12	CF_2Cl_2
1932	R11	$CFCl_3$
1960s	R22	CF_2ClH
1980s	R123	CF_3CCl_2H
1980s	R124	CF_3CFClH
1980s	R125	CF_3CF_2H
1990s	R134a	CF_3CFH_2
1990s	R407C	R32/R125/R134a 23/25/52 wt.%
1990s	R410A	R32/R125 50/50 wt.%
1990s	R404A	R125/R143a/R134a 44/52/4 wt.%

and toxic compounds as well as the inert gases, based on their low boiling points. They were left with just eight elements: carbon, nitrogen, oxygen, sulfur, hydrogen, fluorine, chlorine and bromine. These elements clustered at an intersecting row and column of the periodic table of the elements, with fluorine at the intersection.

Midgley and his colleagues then made three interesting observations. First, flammability decreases from left to right for the eight elements. Second, toxicity generally decreases from the heavy elements at the bottom to the lighter elements at the top. And third, every known refrigerant at the time was made from combinations of those elements. The historical fluids actually comprised only seven of the eight Midgley elements, since there appears to be no record of prior use of refrigerants containing fluorine.

The first publication on fluorochemical refrigerants shows how chlorination and fluorination of hydrocarbons can be varied to provide desired boiling points (Midgley, 1930). This paper also shows how the composition of the molecule influences relative flammability and toxicity. Commercial production of R12 began in 1931, followed by R11 in 1932 (Downing, 1966 and 1984).

Other investigators have repeated Midgley's search with newer methods and modern databases and have come to similar findings. McLinden and Didion (1987) documented an extensive screening of industrial chemicals. Of the chemicals meeting their criteria, all but two — both highly reactive and toxic — contained the Midgley elements.

1.4 REQUIREMENTS FOR WORKING FLUIDS

Refrigerants are expected to meet certain criteria in order to be found acceptable by industry and users. However, with increasing environmental concerns, the set of criteria increases. It becomes more and more difficult to find fluids that meet all or even most requirements. Future refrigerant choices will always be a compromise. The following criteria should be fulfilled:

- *Lack of corrosion, chemically inert.* Refrigerants should not corrode any construction materials present in the system. Also, they should not react with any fluids in the system (such as compressor lubricants).
- *Lack of toxicity.* For the safety of manufacturing and service personnel, refrigerants must be nontoxic.
- *Nonflammable.* To avoid accidents caused by any sources of intense heat or ignition (such as propane torches used by repair personnel or sparks generated by electrical equipment), it is generally required that refrigerants are nonflammable and not explosive except when the quantities are below prescribed limits.
- *Environmentally safe.* Refrigerants should not affect or alter environmental conditions in any way. One important aspect of that requirement is that refrigerants should show high energetic efficiency. The lack thereof would lead to increased production of carbon dioxide at the power plant and thus contribute indirectly to the global warming process.
- *Thermodynamic requirements.* The thermodynamic properties of refrigerants have to fulfill certain criteria that will be discussed in Chapter 2. The overall goal is to achieve energy conversion efficiency and sufficient capacity.

Thus, refrigerants must be safe locally and globally. To be safe locally means fluids should be nontoxic and nonflammable and pose no danger of explosion. They must also not affect any construction materials. These criteria aim at safety during operation, manufacture, service, use and final disposal. To be safe globally means refrigerants should not affect the environment in a significantly negative way.

In order to meet these requirements, an increasing effort is made to use refrigerant mixtures. They offer the potential to be the best compromise for fluids that meet some of the thermodynamic requirements that are very important for the performance of the equipment and hopefully all the other requirements that determine feasibility in many different aspects. Refrigerant mixtures are gaining importance as the acceptable working fluid of future generations of equipment. However, the thermo-physical properties of mixtures are considerably more complex than those of pure fluids. It is the focus of this text to comprehensively present the characteristics of mixtures to the advanced student and practicing engineer.

1.5 BACKGROUND OF ENVIRONMENTAL CONCERNS

1.5.1 IMPACT OF OZONE DEPLETION

Stratospheric ozone depletion and global warming threaten to become the dominant environmental issues. *Chlorofluorocarbons* (CFCs) and other ozone-depleting substances that leak from refrigeration and air-conditioning equipment migrate to the stratosphere and deplete the ozone layer. Current ozone depletion at mid-latitudes is estimated as approximately 5%. Ozone depletion harms living creatures on Earth, increases the incidence of skin cancer and cataracts and poses risks to the human immune system. A sustained 1% decrease in stratospheric ozone will result in approximately a 2% increase in the incidence of fatal non-melanoma skin cancer based on a United Nations Environment Program study. The U.S. Environmental Protection Agency (EPA) expects 295 million fewer cases of non-melanoma skin cancer over the next century with a successful phaseout of CFCs.

1.5.2 IMPACT OF GLOBAL WARMING

The anthropogenic use of energy has been adding gases to the atmosphere that trap heat radiation and warm the earth, known as "greenhouse gases." In 1995, the Intergovernmental Panel on Climate Change (IPCC) reported that the biosphere has already warmed about 0.6°C over the last century and the temperature will increase another 1.1 to 3.6°C in the next century due to a discernible human influence on a global climate. The temperature rise of this magnitude will change local and global climates, temperatures and precipitation patterns, induce a sea level rise and alter the distribution of fresh water supplies. The impact on our health by global warming is likely to be significant.

There are two types of global warming contributions through refrigeration and air-conditioning systems. The first one is the *direct global warming potential* (DGWP) due to the emission of refrigerants and their interaction with heat radiation. The second is the *indirect global warming potential* (IDGWP) due to the emission of CO_2 by consuming the energy that is generated through the combustion of fossil fuels. The combined effect of these two global warming contributions is called the *total equivalent warming impact* (TEWI) (Fish et al., 1991).

TABLE 1.2
HCFCs Regulation Schedule

Jan. 1, 1996	2004	2010	2015	2020	2030
Freeze consumption	35% reduction	65% reduction	90% reduction	99.5% reduction	100% reduction

The base level is the sum of its 1989 ozone-depletion potential (ODP) weighted HCFCs consumption plus 3.1% of its 1989 ODP weighted CFCs consumption.

1.5.3 INTERNATIONAL EFFORTS ON ENVIRONMENTAL PROTECTION

Concern over the potential environmental impacts of ozone depletion led to the development of an international agreement, the Montreal Protocol, to reduce the production of ozone-depleting substances such as CFCs, hydrochlorofluorocarbons (HCFCs) and halons. After the Montreal Protocol was signed in 1987, the regulation was extended in a follow-up conference. At the fourth meeting of the parties to the Montreal Protocol in November 1992, new controls were required to phase out CFCs by the end of 1995 and HCFCs by 2030 (Table 1.2; Reed, 1993). Now the regulation of HCFCs is being tightened with a faster schedule and some countries already have more severe regulation plans. In the United States, the phaseout of R22 in new machinery is set for the year 2010 (Allied Signal Chemicals, 1999) and in Germany it was set for January 1, 2000 (Kruse, 1993).

In 1997, the Parties to the United Nations Framework Convention on Climate Change agreed to an historic Kyoto Protocol to reduce greenhouse gas emissions and set emission reduction targets for developed nations: 8% below 1990 emissions levels for the European Union, 7% for the United States and 6% for Japan. Emission reduction targets include hydrofluorocarbons (HFCs) that are some of the alternative refrigerants introduced as a response to ozone depletion.

1.5.4 IMPACT OF ENVIRONMENTAL ISSUES ON U.S. CLIMATE CONTROL INDUSTRY

The Montreal Protocol and Kyoto Protocol forced the climate control industry to change refrigerants. In the U.S., the Clean Air Act Amendments of 1990 based on the Montreal Protocol forced the climate control industry to change its use of CFCs to HFCs. However, the European Union encourages the use of natural refrigerants such as hydrocarbons and carbon dioxide. Therefore, the search for the alternative refrigerants is still wide open but has to be completed in a limited time to satisfy these two protocols.

1.5.5 MIXTURES AS R22 REPLACEMENTS

R22 is widely used in the air-conditioning and heat pump industry, especially in residential unitary and air-conditioning systems. The phaseout of R22 requires manufacturers to find suitable alternatives in a relatively short time frame. In 1991, the

Air-conditioning and Refrigeration Institute (ARI) started the R22 Alternative Refrigerants Evaluation Program (AREP) to find and evaluate promising alternatives to the refrigerants R22 and R502 for products such as unitary air-conditioners, heat pumps, chillers, refrigeration equipment and ice-making machines (Godwin, 1993). This test program involved researchers worldwide and shared experimental results. Alternative refrigerants were sought to satisfy the following requirements: environmentally benign (zero ODP), nonflammable and similar system level behavior (equivalent performance, minimum hardware changes). Currently there is no acceptable pure fluid replacement that satisfies all these requirements. Alternatively, the mixing of these refrigerants was suggested to fulfill the desired requirements.

REFERENCES

Albritton, D., 1997, "Ozone Depletion and Global Warming," ASHRAE/NIST Refrigerants Conference: Refrigerants for the 21st Century, ASHRAE. Atlanta, GA, pp. 1–5.

Allied Signal Chemicals, 1999, Genetron AZ-20 Product Brochure.

Angelino, G. and C. Invernizzi, 1988, "General method for the thermodynamic evaluation of heat pump working fluids," *International Journal of Refrigeration*, Vol. 11, No. 1, pp. 16–25.

ANSI/ASHRAE Standard 15-1994: "Safety Code for Mechanical Refrigeration," American Society of Heating, Refrigerating, and Air-Conditioning Engineers. Atlanta, GA.

ANSI/ASHRAE Standard 34-1997: "Designation and Safety Classification of Refrigerants (including addenda 34a through 34f)," American Society of Heating, Refrigerating, and Air-Conditioning Engineers, Atlanta, GA.

ARI, 1992, "Centrifugal and Rotary Screw Water-Chilling Packages," Standard 550-92, Air-Conditioning and Refrigeration Institute, Arlington, VA.

ARI, 1996, "Inputs for AFEAS/DOE Phase 3 Study of Energy and Global Warming Impacts," Air-Conditioning and Refrigeration Institute, Arlington, VA.

ARI, 2000, "Statistical Release," Air-Conditioning and Refrigeration Institute, Arlington, VA.

ASHRAE Handbook of Fundamentals, 1993, Chapter 16, American Society of Heating, Refrigerating, and Air-Conditioning Engineers, pp. 2–6.

ASHRAE, 1993, "Addendum to Number Designation and Safety Classification of Refrigerants," ANSI/ASHRAE Standard 34a-1993 (Addendum to ANSI/ASHRAE 34-1992), American Society of Heating, Refrigerating, and Air-Conditioning Engineers, Atlanta, GA.

Atwood, T. and H.M. Hughes, 1990, "Refrigerants and energy efficiency," Proceedings of the 1990 USNC/IIR-Purdue Refrigeration Conference and ASHRAE-Purdue CFC Conference, West Lafayette, IN, pp. 80–89, July.

Bhatti, M.S., 1999, "A Historical Look at Chlorofluorocarbon Refrigerants," *ASHRAE Transactions*, Vol. 105, Pt. 1, pp. 1186–1208.

Bivens, D.B. and B.H. Minor, 1997, "Fluoroethers and Other Next Generation Fluids," ASHRAE/NIST Refrigerants Conference: Refrigerants for the 21st Century, The National Institute of Standards and Technology, Gaithersburg, MD, pp. 119–131.

Bivens, D.B. and, B.H. Minor, 1998, "Fluoroethers and Other Next Generation Fluids," *International Journal of Refrigeration*, Vol. 21, No. 7, pp. 567–576.

Calm, J.M., 1993, "Comparative Global Warming Impacts of Electric Vapor-Compression and Direct-Fired Absorption Equipment," Report TR-103297, Electric Power Research Institute (EPRI), Palo Alto, CA.

Calm, J.M, 1994, "Refrigerant Safety," *ASHRAE Journal*, Vol. 36, No. 7, pp. 17–26.

Calm, J.M., D.J. Wuebbles, and A. Jain, 1997, "Impacts on global ozone and climate from 2,2-dichloro-1,1,1-trifluoroethane (HCFC-123) emissions," *Journal of Climatic Change*, submitted, Car Lighting and Power Company (CLPC), 1922, Advertisement, Ice and Refrigeration, p. 28.

Calm, J.M. and D.A. Didion, 1997, "Trade-Offs in Refrigerant Selections: Past, Present, Future," *International Journal of Refrigeration*, Vol. 21, No. 5, pp. 6–19.

Carrier, W.H. and R.W. Waterfill, 1924, "Comparison of Thermodynamic Characteristics of Various Refrigerating Fluids," *Refrigerating Engineering*.

Clayton, J.W., Jr., 1967, "Fluorocarbon Toxicity and Biological Action," *Fluorine Chemistry Reviews*, Vol. 1, pp. 197–252.

CLPC (Car Lighting and Power Company), 1922, *Ice and Refrigeration*, Advertisement, Nov., p. 28.

Corr, S., J.D. Morrison, F.T. Murphy and R.L. Powell, 1995, "Developing the hydrofluoro-carbon range: Fluids for centrifugal compressors," Proceedings of the 19th International Congress of Refrigeration, Paris, France: International Institute of Refrigeration (IIR), IVa, pp. 31–138.

Dekleva, T.W., 1994, "Flammability Testing: Observations related to HFC systems," Presentation 2.5, ARI Flammability Workshop — Summary and Proceedings, Arlington, VA: Air-Conditioning and Refrigeration Institute (ARI).

Didion, D.A., 1994, "The Impact of Ozone-Safe Refrigerants on Refrigeration Machinery Performance and Operation," *Transactions of the Society of Naval Architects and Engineers*.

Didion, D.A., 1999, "The Application of HFCs as Refrigerants," Proceedings of the Centenary Conference of the Institute of Refrigeration, London, SW1H 9JJ.

Didion, D.A., 1999, "The Influence of the Thermophysical Properties of the New Ozone-Safe Refrigerants on Performance," *International Journal of Applied Thermodynamics*, Vol. 2, No. 1, pp. 19–35.

Didion, D.A. and M.S. Kim, 1999, "NIST Leak and Recharge Simulation Program for Refrigerant Mixtures," NIST Standard Reference Database 73: REFLEAK, Version 2.0, National Institute of Standards and Technology, Gaithersburg, MD.

Dittus, F.W. and L.K.M. Boelter, 1930, University of Calif., Berkeley, *Publ. Engr.*, Vol. 2, p. 443.

Doerr, R.G., D. Lambert, R. Schafer and D. Steinke, 1993, "Stability studies of E-245 Fluoroether, CH_3–CH_2–O–CHF_2," *ASHRAE Transactions*, Vol. 99, Pt. 2, pp. 1137–1140.

Domanski, P.A., D.A. Didion and J.P. Doyle, 1994, "Evaluation of Suction Line — Liquid Line Heat Exchange in the Refrigeration Cycle," *International Journal of Refrigeration*, Vol. 17, No. 7, pp. 487–493.

Domanski, P.A., D.A. Didion and J.S.W. Chi, 1997, "CYCLE_D: NIST Vapor-Compression Design Program," Standard Reference Database 49, Gaithersburg, MD: National Institute of Standards and Technology (NIST).

Domanski, P., 1997, "Theoretical Evaluation of the Vapor Compression Cycle With a Liquid-Line/Suction-Line Heat Exchanger, Economizer, and Ejector," NISTIR 5606, The National Institute of Standards and Technology, Gaithersburg, MD.

Downing, R.C., 1966, "History of the organic fluorine industry," *Kirk-Othmer Encyclopedia of Chemical Technology* (2nd edition). New York, NY: John Wiley & Sons, Incorporated, Vol. 9, pp. 704–707.

Downing, R.C., 1984, "Development of chlorofluorocarbon refrigerants," *ASHRAE Transactions*. Atlanta, GA: American Society of Heating, Refrigerating, and Air-Conditioning Engineers (ASHRAE), Vol. 90, Pt. 2B, pp. 481–491.

Downing, R.C., 1988, *Fluorocarbon Refrigerants Handbook*, Englewood Cliffs, NJ, Prentice Hall.

DuPont Fluorochemicals, 2000, "Technical Information for Suva Refrigerants," http://www. dupont.com/suva, Wilmington, DE.

Evans, O., 1805, *The Abortion of a Young Steam Engineer's Guide*, Philadelphia, PA.

Fairchild, P. et al., 1991, "Total Equivalent Global Warming Impact: Combining Energy and Fluorocarbon Emissions Effects," Proceedings of the International CFC and Halon Alternatives Conference, Baltimore, MD.

Fischer, S.K., P. Hughes and P. Fairchild, 1991, "Energy and Global Warming Impacts of CFC Alternative Technologies," Alternative Fluorocarbons Environmental Acceptability Study (AFEAS) and U.S. Department of Energy (DOE).

Fischer, S.K., J.J. Tomlinson and P.J. Hughes, 1994, "Energy and Global Warming Impacts of Not-In-Kind and Next Generation CFC and HCFC Alternatives," Alternative Fluorocarbons Environmental Acceptability Study (AFEAS) and U.S. Department of Energy (DOE).

Glamm, P.R., E.F. Keuper and F.B. Hamm, 1996, "Evaluation of HFC-245ca for Commercial Use in Low Pressure Chillers," Report DOE/CE/23810-67, Arlington, VA, Air-Conditioning and Refrigeration Technology Institute (ARTI).

Godwin, D., 1993. "Results of System Drop-In Tests in ARI's R-22 Alternative Refrigerants Evaluation Program, Arlington, VA: ARI.

Gorenflo, D. and V. Bieling, 1986, "Heat transfer in pool boiling of mixtures with R-22 and R-115," XVII Int. Symposium on Heat and Mass Transfer in Cryogenics and Refrigeration, International Institute of Refrigeration, Paris.

Houghton, L.G. et al., 1995, "Climate Change 1994," Radiative Forcing of Climate Change and an Evaluation of the IPCC IS92 Emissions Scenarios, Cambridge University Press.

Huber, M.L., J.S. Gallagher, M.O. McLinden and G. Morrison, 1996, "NIST Thermodynamic Properties of Refrigerants and Refrigerant Mixtures Database (REFPROP)," Standard Reference Database 23, Version 5.0. Gaithersburg, MD, National Institute of Standards and Technology (NIST).

Ingels, M., 1952, *Willis Haviland Carrier — Father of Air Conditioning*. Syracuse, NY: Carrier Corporation.

Intergovernmental Panel on Climate Change (IPCC), 1996, "Climate Change 1995 — Contribution of Working Group I to the Second Assessment Report of the Intergovernmental Panel on Climate Change," edited by J.T. Houghton, L.G. Meira Filho, B.A. Callander, N. Harris, A. Kattenberg and K. Maskell. Cambridge, UK: Cambridge University Press.

Johnson, J. and T. Watson, 1993, "R-134A Compatibility with Centrifugal Chillers," Proceedings of the International Seminar on New Technology of Alternative Refrigerants, Keidanren Hall, Tokyo, Japan.

Kates, R.W. et al., National Academy Report, 1999, "Our Common Journey, a Transition Toward Sustainability," National Academy Press, Washington D.C., USA.

Kaul, M., M. Kedzierski and D.A. Didion, 1996, "Horizontal Flow Boiling of Alternative Refrigerants within a Fluid Heated Micro-Fin Tube," Process, Enhanced, and Multiphase Heat Transfer: A Festschrift for A.E. Bergles, Begell House, Inc., New York, NY.

Kedzierski, M.A., J.H. Kim and D.A. Didion, 1992, "Causes of the Apparent Heat Transfer Degradation for Refrigerant Mixtures," ASME HTD-197, Two Phase Flow and Heat Transfer, American Society of Mechanical Engineers, New York, NY.

Kim, Man-Hoe, P.A. Domanski and D.A. Didion, 1997, "Performance of R-22 Alternative Refrigerants in a System with Cross-Flow and Counter-Flow Heat Exchangers," NISTIR 5945, National Institute of Standards and Technology, Gaithersburg, MD.

Kruse, H., 1993, "European Research and Development Concerning CFC and HCFC Substitution," Proceedings of ASHRAE/NIST Refrigerants Conference, Gaithersburg, MD, pp. 41–57.

Lorenz, H., 1984, *Die Ausnutzung der Brennstoffe in den Kühlmaschinen* [Use of Flammable Substances (commonly Fuels) in Refrigeration Equipment]. Germany: Zeitschrift für die gesamte Kälte Industrie, Vol. 1, pp. 10–15.

Marques, M., S.A. Multibras and P.A. Domanski, 1998, "Potential Coefficient of Performance Improvements due to Glide Matching with R407C," Seventh International Refrigeration Conference at Purdue, West Lafayette, IN, pp. 101–108.

McLinden, M.O. and D.A. Didion, 1987, "Quest for alternatives," *ASHRAE Journal*, Vol. 29, No. 12, pp. 32–36, 38, 40 and 42.

McLinden, M.O., 1988, "Thermodynamic Evaluation of Refrigerants in the Vapor Compression Cycle Using Reduced Properties," *International Journal Refrigeration*, Vol. 11, No. 3, pp. 134–143.

McLinden, M.O., 1990, "Optimum refrigerants for non-ideal cycles — An analysis employing corresponding states," Proceedings of the 1990 USNC-IIR-Purdue Refrigeration Conference and ASHRAE-Purdue CFC Conference, West Lafayette, IN, pp. 69–79.

McLinden, M., S.A. Klein, E.K. Lemmon and A.P. Perkin, 1998, "Thermodynamic and Transport Properties of Refrigerants and Refrigerant Mixtures," NIST Database 23: NIST REFPROP, Version 6.0, National Institute of Standards and Technology, Gaithersburg, MD.

Midgley, T., Jr. and A. L. Henne, 1930, "Organic fluorides as refrigerants," *Industrial and Engineering Chemistry*, Vol. 22, pp. 542–545.

Midgley, T., Jr., 1937, "From the periodic table to production," *Industrial and Engineering Chemistry*, Vol. 29, No. 2, pp. 239–244.

Molina, M. J. and F.S. Rowland, 1974, "Stratospheric sink for Chlorofluoromethanes: Chlorine atom catalyzed destruction of ozone," *Nature*, Vol. 249, pp. 810–812.

Morrison, G., 1974, "The shape of the temperature-entropy saturation boundary," *International Journal of Refrigeration*, Vol. 17, No. 7, pp. 494–504.

Nagengast, B.A., 1989, "A history of refrigerants," *CFCs: Time of Transition*, Atlanta, GA, American Society of Heating, Refrigerating, and Air-Conditioning Engineers, pp. 3–15.

Nagengast, B.A., 1996, "History of sealed refrigeration systems," *ASHRAE Journal*, Vol. 38, No. 1, S37, S38, S42–S46 and S48.

NERC, 1995, Electric Supply and Demand Database (Version 2.0), Princeton, NJ: North American Electric Reliability Council.

Perkins, J., 1834, Apparatus for Producing Ice and Cooling Fluids, Patent No. 6662, United Kingdom.

Petterson, J., 1999, "Carbon Dioxide (CO_2) as a Primary Refrigerant," Proceedings of the Centenary Conference of the Institute of Refrigeration, London, SW1H 9JJ.

Pictet, R., 1885, *Aus Kohlensäure und schwefliger Säure bestehende Verflüchtigungsflüssigkeit für Kältemaschinen* [A Volatile Liquid Comprising Carbon Dioxide and Sulfur Dioxide for Refrigeration Machines], Patent 33733 (Class 17, Ice Production and Storage), Germany: Kaiserliches Patentamp.

Pillis, J.W., 1993, "Expanding Ammonia Usage in Air Conditioning," Proceedings of the ASHRAE/NIST Refrigerants Conference: R-22/R-502 Alternatives, American Society of Heating, Refrigerating, and Air Conditioning Engineers, Atlanta, GA, pp. 103–108.

Reed, J.W., 1993, "Environmental Overview: CFC and HCFC Regulatory Update," Proceedings of the 4th IEA Heat Pump Conference, Maastricht, the Netherlands, pp. 11–19.

Richard, R.G. and I.R. Shankland, 1992, "Flammability of alternative refrigerants," *ASHRAE Journal*, Vol. 34, No. 4, pp. 20 and 22–24, April.

Ross, H., R. Radermacher, M. di Marzo and D. Didion, 1987, "Horizontal flow boiling of pure and mixed refrigerants," *International Journal of Heat and Mass Transfer*, Vol. 30, pp. 972–992.

Sand, J.R. and S.K. Fischer, 1994, "Modeled Performance of Non-chlorinated Substitutes for CFC-11 and CFC-12 in Centrifugal Chillers," *International Journal of Refrigeration*, Vol. 17, No. 1, pp. 40–48.

Sand, J.R., S.K. Fischer and V.D. Baxter, 1997, "Energy and Global Warming Impacts of HFC Refrigerants and Emerging Technologies," Alternative Fluorocarbons Environmental Acceptability Study (AFEAS) and U.S. Department of Energy (DOE).

Smith, N.D., K. Ratanaphruks, M.W. Tufts and A.S. Ng, 1993, "R-245ca: A potential far-term alternative for R-11," *ASHRAE Journal*, Vol. 35, No. 2, pp. 19–23.

Smith, N.D. and M.W. Tufts, 1994, "Flammable properties of HFC refrigerants — Some fundamental considerations," Presentation 2.2, ARI Flammability Workshop — Summary and Proceedings, Arlington, VA: Air-Conditioning and Refrigeration Institute (ARI).

Starner, K.E., 1993, "Heat Exchangers for Ammonia Water Chillers: Design Considerations and Research Needs," Proceedings of the ASHRAE/NIST Refrigerants Conference: R-22/R-502 Alternatives, American Society of Heating, Refrigerating, and Air Conditioning Engineers, Atlanta, GA, pp. 85–90.

Stera, A.C., 1993, "Developments in Transportation of Chilled Produce by Sea and Air," IIR Conference Proceedings: Cold Chain Refrigeration Equipment by Design, International Institute of Refrigeration, Paris.

Stoecker, W., 1989, "Expanded Applications for Ammonia — Coping with Releases to the Atmosphere, CFCs: Today's Options — Tomorrow's Solutions," Proceedings of ASHRAE's 1989 CFC Technology Conference, ASHRAE, Atlanta, GA.

Thévenot, R., 1979, *A History of Refrigeration throughout the World*, Translated from French by J.C. Fidler. Paris, France: International Institute of Refrigeration.

Threlkeld, J.L., 1970, *Thermal Environmental Engineering* (2nd edition), Prentice Hall, Inc.

UNEP, 1987, "Montreal Protocol on Substances That Deplete the Ozone Layer," United Nations Environment Programme (UNEP).

UNEP, 1996, *Handbook for the Montreal Protocol on Substances That Deplete the Ozone Layer* (4th edition), Nairobi, Kenya: United Nations Environment Programme (UNEP) Ozone Secretariat.

WMO, 1994, "Scientific Assessment of Ozone Depletion: 1994," Chaired by D.L. Albritton, R.T. Watson and P.J. Aucamp, Global Ozone Research and Monitoring Project Report 37, Geneva, Switzerland, World Meteorological Organization (WMO).

Wood, B.D., 1982, *Applications of Thermodynamics* (2nd edition). Reading, MA, Addison-Wesley Publishing Company, p. 40.

Wuebbles, D.J., 1981, "The Relative Efficiency of a Number of Halocarbons for Destroying Stratospheric Ozone," Report UCID-18924, Livermore, CA, Lawrence Livermore National Laboratory (LLNL).

Wuebbles, D.J., 1995, "Weighing functions for ozone depletion and greenhouse gas effects on climate," *Annual Review of Energy and Environment*, Vol. 20, pp. 45–70.

Wuebbles, D.J. and J.M. Calm, 1997, "An environmental rationale for retention of endangered chemical species," *Science*.

2 Properties of Working Fluids

The performance and efficiency of reversible cycles are independent of the properties of any working fluids. However, the performance and efficiency of a real machine are determined to a large degree by the properties of the working fluids. Furthermore, both the first cost and the operating cost of a heat pump are strongly dependent on the refrigerant properties.

2.1 THERMODYNAMIC DIAGRAMS OF PURE AND MIXED REFRIGERANTS

This chapter describes the graphic representations of the properties of pure and mixed refrigerants. Graphical representations foster the development of a good intuitive understanding of important characteristics of working fluids while helping to clarify cycle behavior. Some conventional pure fluids, selected because of their favorable environmental characteristics, are included here for comparison purposes. In the following, typical diagrams of fluid properties are discussed. In addition, the determination of the properties for cycle calculations is explained in detail.

When designing a heat pump, the most important thermodynamic variables to be considered are: pressure, temperature, mass fraction (for mixtures), enthalpy, specific volume and entropy. To display all variables, a multidimensional diagram is required. This is not practical, so several two-dimensional diagrams are commonly used. These diagrams show any two variables on their axes and display other variables as sets of curves of constant properties such as isobars and isotherms. Usually, T-s, $\ln(P)$-h or h-s diagrams are used for design calculations of cycles with pure fluids. However, when mixtures are used, additional variables and concentrations must be considered. In some cases, this is accomplished by showing the conventional diagrams for constant concentration. Thus, new versions of the same diagram must be generated when a different concentration of the same mixture is of interest. Other diagrams show concentration as the independent variable. These are uniquely suited to study the characteristics of mixtures.

For the sake of simplicity, the explanation focuses on pure fluids and two-component (binary) mixtures only. From a thermodynamic point of view, a two-component mixture possesses one additional degree of freedom as compared to a pure fluid — the concentration. The *concentration* of a mixture may be defined in many different ways including mole fraction, mass fraction and the like. In this text, the *mass fraction* is used throughout, defined as

$$x = \frac{\text{mass of one component [kg]}}{\text{mass of all components [kg]}} \tag{2.1}$$

17

This variable includes values in the range $0.0 \leq x \leq 1.0$, with 0.0 and 1.0 denoting the respective pure components. The various definitions for variables describing mixture compositions are discussed in Section 2.2.2.

2.1.1 TEMPERATURE-MASS FRACTION DIAGRAM

We begin with one of the diagrams that is most important for the understanding of working fluid mixtures, the temperature-mass fraction (*T-x*) diagram (Figure 2.1.1).

When the liquid and vapor phases of a mixture coexist in equilibrium, the saturation temperature varies with the mass fraction even though the pressure is constant. This behavior is in contrast to a pure fluid where T_{sat} remains constant. Figure 2.1.1 shows the temperature-mass fraction diagram (*T-x* diagram) for a mixture of two components, R134a and R32, at constant pressure of 1.0 MPa. The mass fraction axis ranges from 0.0 (pure R134a) to 1.0 (pure R32). The area below the bubble line represents subcooled liquid. The area above the dew line represents superheated vapor. The area enclosed by the bubble and dew lines is the two-phase range region. The boiling point for a mixture of mass fraction *x* is located on the bubble line at that mass fraction. The bubble line indicates the saturation temperature at which the first vapor bubble is formed for the specified pressure and mass fraction when the temperature is raised from the subcooled liquid region to the saturation value.

The boiling points of the pure components T_{R134a} and T_{R32} are found on the respective ordinates. According to Figure 2.1.1, the boiling point of R134a is higher than that of R32. The dew line indicates the temperature at which the first liquid droplet is formed when a superheated gas mixture of a given mass fraction is cooled at constant pressure.

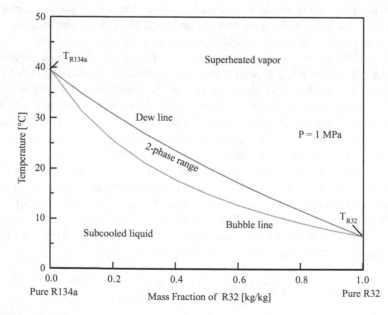

FIGURE 2.1.1 Temperature-mass fraction diagram for R32/R134a.

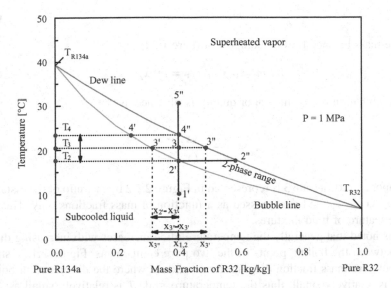

FIGURE 2.1.2 Evaporation process of R32/R134a.

To demonstrate the use of the diagram, a slow (equilibrium) evaporation process at constant pressure is discussed as an example in Figure 2.1.2. The process begins with subcooled liquid at point 1′. Points with a single prime denote the liquid phase; points with a double prime, the vapor phase. As the mixture is heated, the temperature increases and the bubble line is reached. This is point 2′. Here the first vapor bubble forms. The mass fraction of the first vapor bubble is found at point 2″. This vapor is in thermal equilibrium with the liquid phase at point 2′. The vapor is enriched in R32 as compared to the liquid and has a mass fraction of x_2. This is a consequence of the fact that at the same temperature, R32 has a higher vapor pressure than R134a. Thus, the vapor phase is bound to contain more R32 molecules.

As the heating process continues, the evaporation process proceeds to point 3, where the mass fraction of the vapor in equilibrium with the remaining liquid is represented by point 3″. The mass fraction of the liquid is now indicated by point 3′. At this point, the amount of R32 in the remaining liquid has been reduced as compared to point 2′ while the vapor is enriched in R32. However, the vapor contains a lower fraction of R32 and more of R134a than at point 2′. The same is true for the remaining liquid. As the evaporation process proceeds, the state points of the liquid and vapor phases continue to follow the bubble and dew point lines. When point 4″ is reached, the evaporation process is complete. The vapor has the same mass fraction as the original subcooled liquid and the mass fraction of the last liquid droplet that evaporated is indicated by point 4′. Further heating produces superheated vapor at point 5″.

During the constant pressure evaporation process, the saturation temperature changed from T_2 to T_4. The temperature difference $(T_4 - T_2)$ is termed *temperature glide*. Vapor quality x_q (defined as the ratio of mass of vapor over total mass) at point 3 can be calculated based on a mass balance for the mixture and one pure component. The overall mass balance is given by

$$m_{3''} + m_{3'} = m_{2'} \qquad (2.2)$$

and the mass balance for one component (here R32)

$$m_{3''} x_{3''} + m_{3'} x_{3'} = m_2 x_{2'} \qquad (2.3)$$

After elimination of $m_{3'}$, the vapor quality is obtained as:

$$x_q = \frac{m_{3''}}{m_{2'}} = \frac{(x_{2'} - x_{3'})}{(x_{3''} - x_{3'})} \qquad (2.4)$$

The vapor quality at state 3 is represented in Figure 2.1.2 by the ratio of the distances $(x_3 - x_{3'})/(x_{3''} - x_{3'})$ and is expressed as a function of mass fractions only. This is a unique feature of fluid mixtures.

It is noted that generally the temperature glide increases with increasing difference between the boiling points of the two pure components. Figure 2.1.3a shows a temperature-mass fraction diagram for a mixture where the difference in boiling points is relatively small, thus the temperature glide T is relatively small as well. Figure 2.1.3b shows the temperature-mass fraction diagram of a mixture where the difference in the boiling points is large. Accordingly, the temperature glide marked as T is large as well. The size of the temperature glide is also a function of the mass fraction. For small and large x, the glide is generally smaller than for intermediate values of x. The mixtures that show the same characteristics as R32/R134a in Figures 2.1.1 through 2.1.3 are traditionally termed nonazeotropic mixtures or, in more recent literature, *zeotropic mixtures*. The name implies that in phase equilibrium, the mass fractions of the vapor and liquid phases are always different.

Some fluids form from azeotropic mixtures, such as a mixture of R12 and R152a, or a mixture of water and ethanol. For an azeotropic mixture, the mass fractions of the liquid and vapor phase are identical at a certain pressure and temperature as shown in Figure 2.1.4. This state is called the *azeotropic point* and the mass fraction of this point is called the *azeotropic mass fraction*. The temperature glide is zero at this point. At all other mass fractions, the mixture exhibits zeotropic behavior. The difference in mass fraction between the liquid and vapor phases changes its sign when the overall mass fraction varies from a value less than the azeotropic mass fraction to a value larger than the azeotropic mass fraction.

There are unique features of azeotropic mixtures that can make their use beneficial. First, they provide another unique normal boiling point that might not be covered by one of the pure compound refrigerants and therefore the possibility exists of fitting some specific application more precisely. Second, they have all of the benefits of pure compound refrigerants — constant boiling point and heat of vaporization, without the fractionation. Third, they can make up for the physical or chemical deficiency of one of the zeotropic components, like covering up the flammability of R152a in R500 (Fairchild, 2000).

There are two types of azeotropes, distinguished from one another by the location of the boiling points at the azeotropic mass fraction relative to the boiling points of the pure fluids. The boiling point can be either higher than the boiling point of both

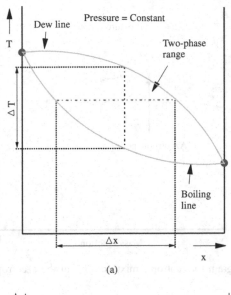

(a)

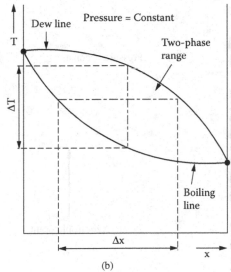

(b)

FIGURE 2.1.3 Temperature-mass fraction diagrams with constituents of varying differences in boiling points.

of the two constituents of the mixture, as in Figure 2.1.4, or lower than the boiling point of the two constituents, as illustrated in Figure 2.1.5.

Figure 2.1.6 shows the *T-x* diagram for the azeotrope R507A at a pressure of 0.1 MPa. R507A is a mixture of R125 and R143a with a mass fraction of 0.5 of R125. Although Figure 2.1.6 shows that the mass fraction of the azeotropic point is approximately 0.35 of R125 (the maximum in the saturation line), it is also noteworthy that the temperature glide is very small throughout the entire mass fraction range.

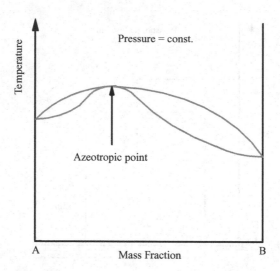

FIGURE 2.1.4 *T-x* diagram for zeotropic mixture. (T_{sat} of the azeotropic point $> T_{sat}$ of the pure fluid.)

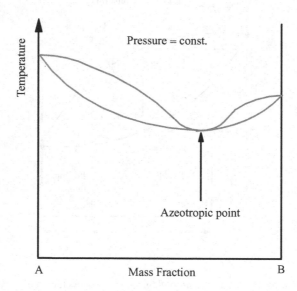

FIGURE 2.1.5 *T-x* diagram for zeotropic mixture. (T_{sat} of the azeotropic point $< T_{sat}$ of the pure fluid.)

This is a consequence of two characteristics: The boiling points of the pure components are very close and there is not much of an opportunity for the bubble and dew line to diverge because the two components form an azeotrope. The actual mass fraction of the azeotropic point is not of much importance for applications in heat pumps; the fact that the temperature glide is negligible is the important characteristic.

R125/R143a Mixture

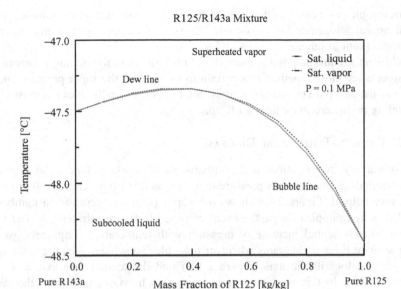

FIGURE 2.1.6 *T-x* diagram for azeotrope R507A (*P* = 0.1 MPa).

R125/R143a Mixture

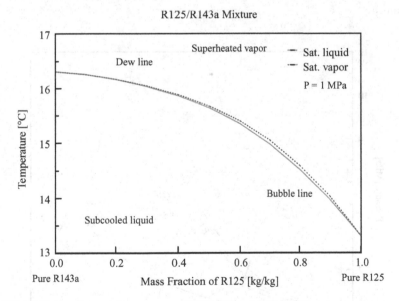

FIGURE 2.1.7 *T-x* diagram for azeotrope R507A (*P* = 1 MPa).

Figure 2.1.7 shows the *T-x* diagram for R507A at a pressure of 1 MPa. An inspection and comparison of Figures 2.1.6 and 2.1.7 reveal that the azeotropic point moves. In this case, the mass fraction of the azeotropic point decreases as the pressure is increased from 0.1 MPa (Figure 2.1.6) to 1 MPa (Figure 2.1.7) and the azeotropic point seems to disappear to the left. This is an interesting characteristic of azeotropes.

The azeotropic point changes its mass fraction with the saturation pressure level and can disappear altogether for certain conditions. A zeotropic mixture may have an azeotropic point at other conditions and vice versa.

While an azeotropic point is more likely to exist when the boiling points of the constituents are close together, it is certain to exist when the vapor pressure curves $P_{sat}(T)$ of the two pure components cross. Such a point usually does not exist when the boiling points are more than 30 K apart.

2.1.2 PRESSURE-TEMPERATURE DIAGRAM

For preliminary investigations and comparisons of working fluids, the pressure-temperature diagram (or vapor pressure diagram, as it is often referred to) turns out to be very helpful. Figure 2.1.8 shows the vapor pressure curves for a number of pure fluids as examples for past, current or possibly future refrigerants. The lines suggest an exponential increase of pressure with temperature. Engineers usually prefer straight lines in diagrams wherever possible. To achieve this, the ordinate is changed to a logarithmic axis (Figure 2.1.9) and the temperature axis to $(-1/T)$ (Figure 2.1.10). In this way, the temperature axis becomes nonlinear, the vapor pressure curves become almost straight lines and the temperature still increases from left to right. The reason for this selection is given by the Clapeyron's Equation described in Section 2.2. This diagram is referred to in this text as the $\ln(P)$-$(-1/T)$ diagram.

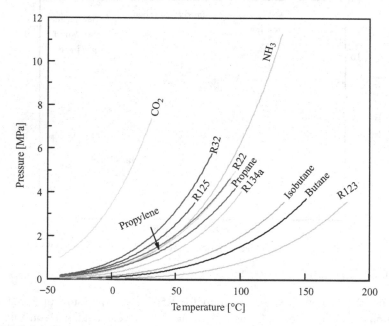

FIGURE 2.1.8 Vapor pressure curves for a number of pure fluids. The curves end at the critical points of the respective fluids.

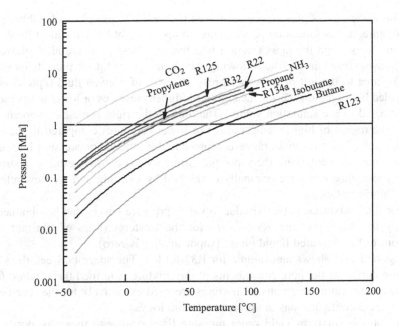

FIGURE 2.1.9 Vapor pressure curves on a ln(P)-T diagram.

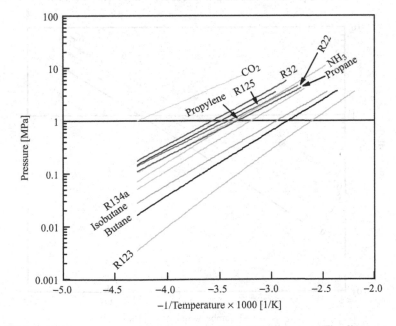

FIGURE 2.1.10 Vapor pressure curves on a ln(P)-(−1/T) diagram. (The lines are almost straight.)

The advantage of this representation is the fact that the plots of bubble point temperature versus saturation pressure are almost straight lines for most fluids and fluid mixtures when the mass fraction is constant. These plots are also referred to as *vapor pressure curves*. The following statements are valid for pure fluids only.

The area to the left of the vapor pressure curve of a given fluid represents the subcooled liquid region. This is the region of higher pressures or lower temperatures as compared to the saturation values. The area to the right represents superheated vapor, the region of higher temperatures or lower pressures compared to the saturation values. The two-phase range is represented by the vapor pressure line itself.

In some representations, the vapor pressure curves are drawn exactly as straight lines. These diagrams were originally developed by Dühring and are now referred to as *Dühring plots*.

For fluid mixtures, curves similar to vapor pressure curves can be obtained by plotting the vapor pressure versus $-1/T$ for the isosteres (lines of constant mass fraction) of the saturated liquid phase (vapor quality is zero).

Figure 2.1.11 shows an example for R32/R134a. The space between the vapor pressure curves of the pure constituents of the mixture is termed the *solution field*. Pressure-temperature diagrams of mixtures are used extensively for the representation of cycle configurations in absorption technology.

Azeotropic mixtures yield vapor pressure lines similar to those for pure components shown in Figure 2.1.11. However, while in zeotropic mixtures all vapor pressure curves for mass fractions x with $0.0 < x < 1.0$ are located between the lines for the two pure components, the vapor pressure curve for the azeotropic mass

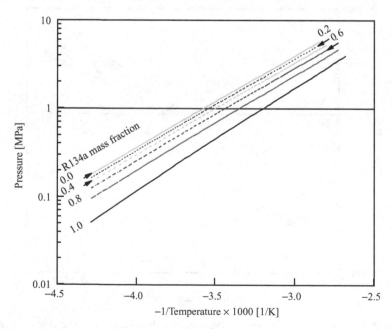

FIGURE 2.1.11 $\ln(P)$-$(-1/T)$ diagram for R32/R134a.

fraction of an azeotropic mixture is located outside the range limited by the vapor pressure curves of the pure components.

2.1.3 Pressure-Enthalpy Diagram

Figures 2.1.12 through 2.1.24 show examples of pressure-enthalpy ($\ln(P)$-h) diagrams for an number of pure fluids and mixtures. This diagram is widely used in the refrigeration field for the evaluation and design of vapor compression cycles.

The diagram and its various parts are explained next. Figure 2.1.25 shows a schematic of the diagram for a pure fluid. The ordinate shows the logarithm of the pressure, the abscissa the specific enthalpy. The two-phase range is found under the *vapor dome*, the area where the isotherms are horizontal. To the left of the two-phase region is the subcooled liquid region and to the right is the superheated vapor region. The boundary line between the subcooled liquid range and the two-phase region is the *saturated liquid line* and the boundary between the superheated vapor and the two-phase range is the *saturated vapor line*. The lines of saturated liquid and saturated vapor converge at the *critical point*. Above the critical point there is no distinction between phases, i.e., no meniscus exists. A $\ln(P)$-h diagram shows several sets of lines of constant properties. Figure 2.1.25 shows only one line as an example for the various sets. The first discussed here is the isotherm.

The *isotherm* of this system shows a slight increasing curvature toward the horizontal (but does not reach a zero slope) as the critical point is approached close to the saturated liquid line. This gives the subcooled enthalpy of a given temperature a slightly lower value than that of saturated liquid at pressures closer to the critical point. R410A is an example where this effect can become significant at higher pressures. In the two-phase region, the isotherm is a straight horizontal line. This is a consequence of the fact that pressure and temperature are constant during a phase change process. In the superheated vapor region, the isotherm curves very quickly to the vertical and approaches ideal gas behavior (i.e., the enthalpy of an ideal gas is independent of the pressure). The change in slope of the isotherm at the transition points between regions is not steady. Above the critical point, the slope of the isotherm is negative until very high pressures are reached; however, the second derivative changes its sign in the neighborhood of the critical point. Only at the critical point itself does the isotherm have a saddle point. This isotherm is sometimes called the *critical isotherm*.

The *isentrope*, the line of constant entropy, is shown frequently in the vapor phase only where it has a positive slope. In Figure 2.1.25, a single isentrope is shown as an example. Also indicated is how the isentrope extends into the two-phase range. There is no abrupt change in the slope of the isentrope at the saturation line. The isentrope is very important to determine compressor work as shown in the text to follow.

The *isochore*, the line of constant volume or density, behaves similarly to the isentrope but shows a much flatter slope. Isochores are important in determining the volumetric capacity of the working fluid.

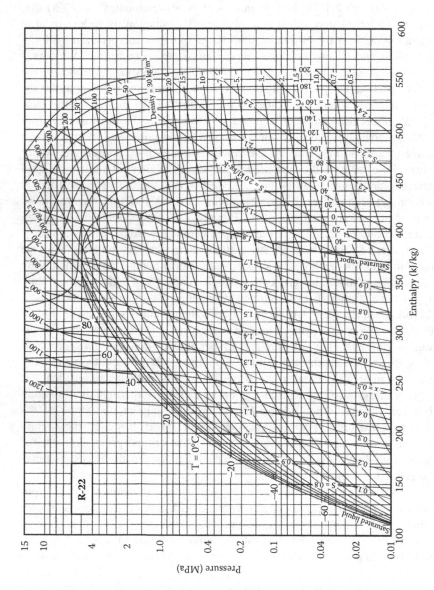

FIGURE 2.1.12 Pressure-enthalpy diagram of R22. (Copyright 2004, American Society of Heating, Refrigerating, and Air-Conditioning Engineers, Inc. Reprinted by permission from ASHRAE 2001 Handbook — Fundamentals.)

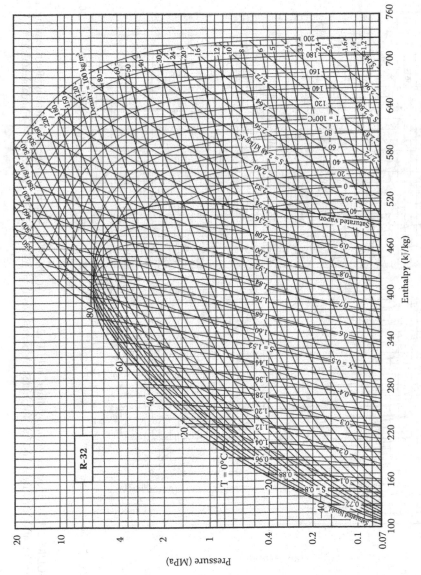

FIGURE 2.1.13 Pressure-enthalpy diagram of R32. (Copyright 2004, American Society of Heating, Refrigerating, and Air-Conditioning Engineers, Inc. Reprinted by permission from ASHRAE 2001 Handbook — Fundamentals.)

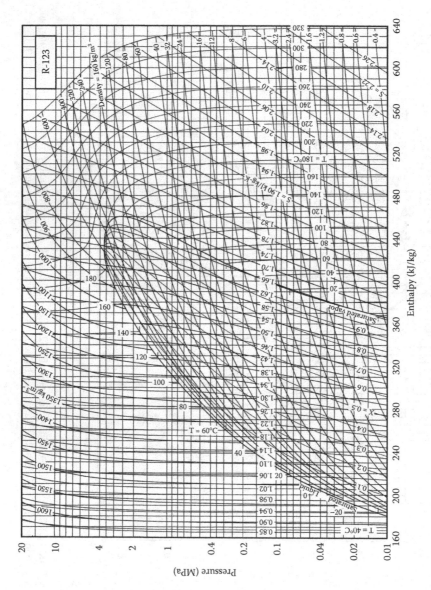

FIGURE 2.1.14 Pressure-enthalpy diagram of R123. (Copyright 2004, American Society of Heating, Refrigerating, and Air-Conditioning Engineers, Inc. Reprinted by permission from ASHRAE 2001 Handbook — Fundamentals.)

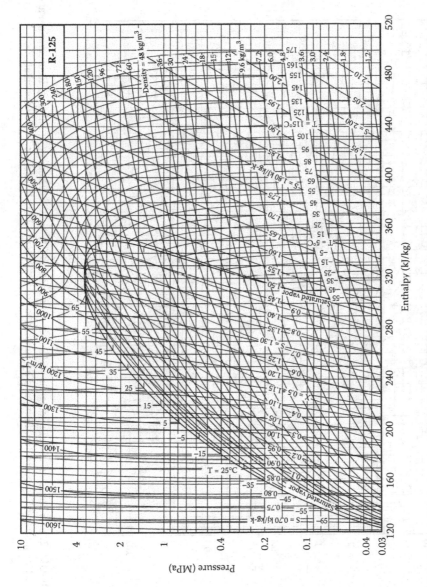

FIGURE 2.1.15 Pressure-enthalpy diagram of R125. (Copyright 2004, American Society of Heating, Refrigerating, and Air-Conditioning Engineers, Inc. Reprinted by permission from ASHRAE 2001 Handbook — Fundamentals.)

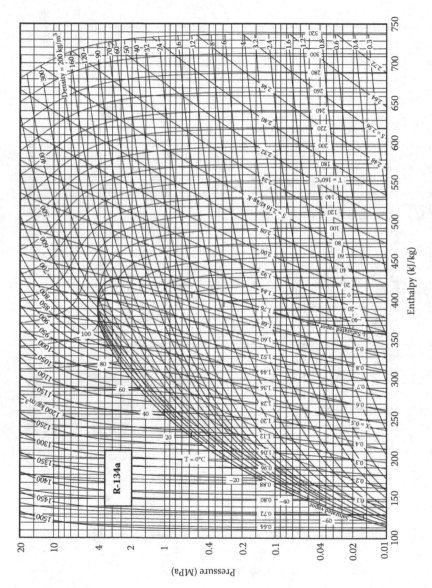

FIGURE 2.1.16 Pressure-enthalpy diagram of R134a. (Copyright 2004, American Society of Heating, Refrigerating, and Air-Conditioning Engineers, Inc. Reprinted by permission from ASHRAE 2001 Handbook — Fundamentals.)

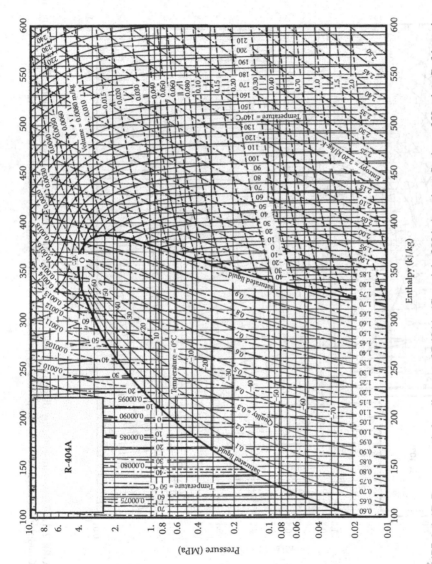

FIGURE 2.1.17 Pressure-enthalpy diagram of R404A. (Copyright 2004, American Society of Heating, Refrigerating, and Air-Conditioning Engineers, Inc. Reprinted by permission from ASHRAE 2001 Handbook — Fundamentals.)

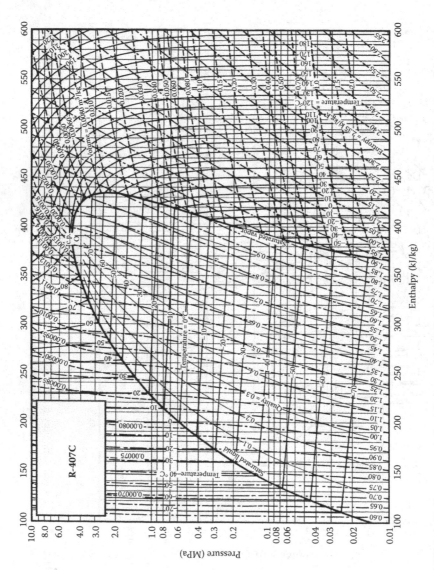

FIGURE 2.1.18 Pressure-enthalpy diagram of R407C. (Copyright 2004, American Society of Heating, Refrigerating, and Air-Conditioning Engineers, Inc. Reprinted by permission from ASHRAE 2001 Handbook — Fundamentals.)

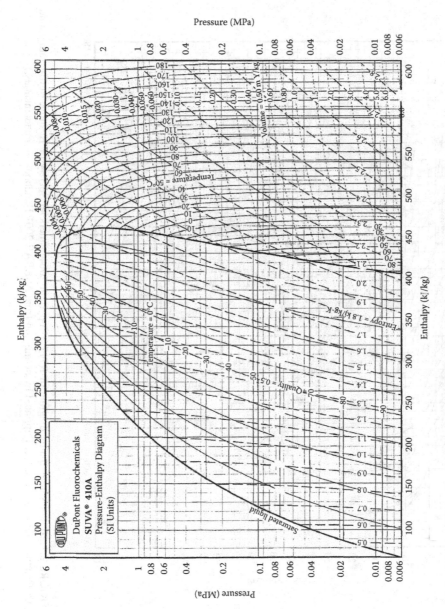

FIGURE 2.1.19 Pressure-enthalpy diagram of R410A. (Chart supplied by DuPont Refrigerants.)

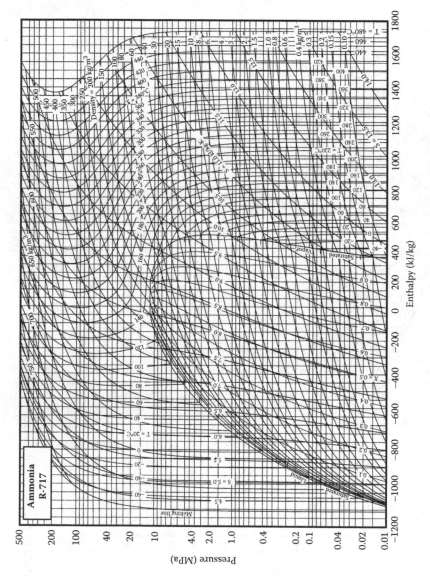

FIGURE 2.1.20 Pressure-enthalpy diagram of R717. (Copyright 2004, American Society of Heating, Refrigerating, and Air-Conditioning Engineers, Inc. Reprinted by permission from ASHRAE 2001 Handbook — Fundamentals.)

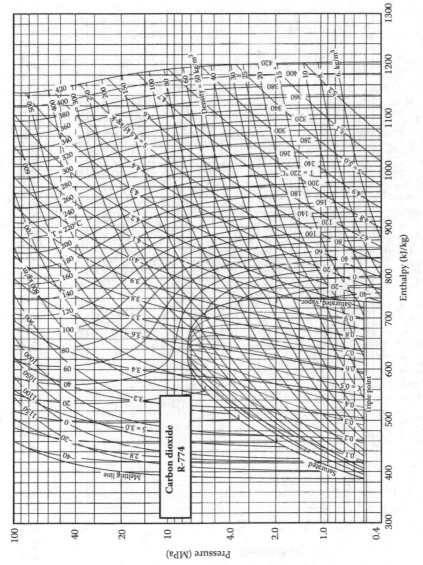

FIGURE 2.1.21 Pressure-enthalpy diagram of R774. (Copyright 2004, American Society of Heating, Refrigerating, and Air-Conditioning Engineers, Inc. Reprinted by permission from ASHRAE 2001 Handbook — Fundamentals.)

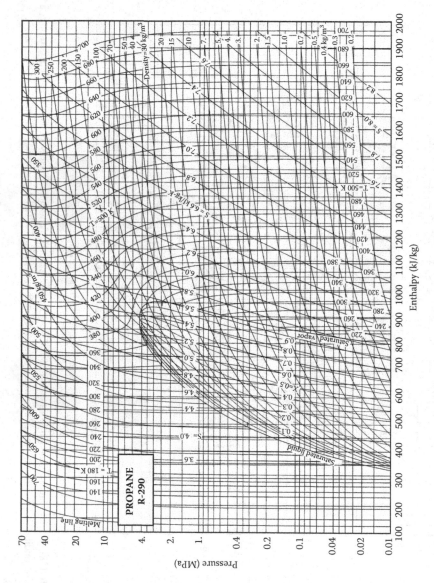

FIGURE 2.1.22 Pressure-enthalpy diagram of R290. (Copyright 2004, American Society of Heating, Refrigerating, and Air-Conditioning Engineers, Inc. Reprinted by permission from ASHRAE 2001 Handbook — Fundamentals.)

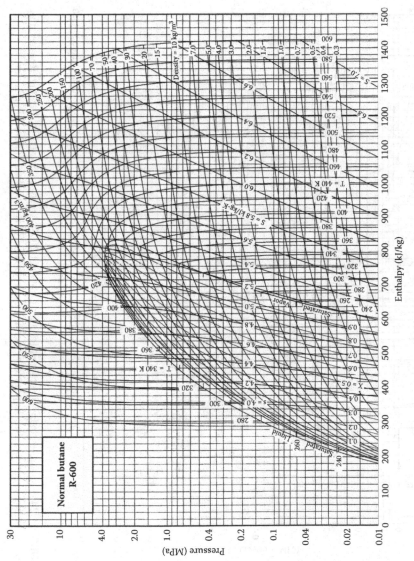

FIGURE 2.1.23 Pressure-enthalpy diagram of R600. (Copyright 2004, American Society of Heating, Refrigerating, and Air-Conditioning Engineers, Inc. Reprinted by permission from ASHRAE 2001 Handbook — Fundamentals.)

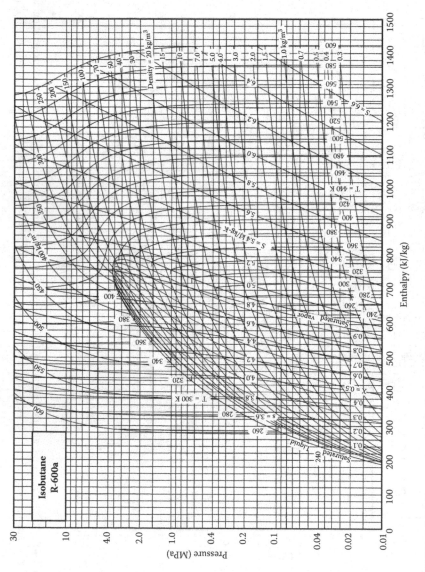

FIGURE 2.1.24 Pressure-enthalpy diagram of R600a. (Copyright 2004, American Society of Heating, Refrigerating, and Air-Conditioning Engineers, Inc. Reprinted by permission from ASHRAE 2001 Handbook — Fundamentals.)

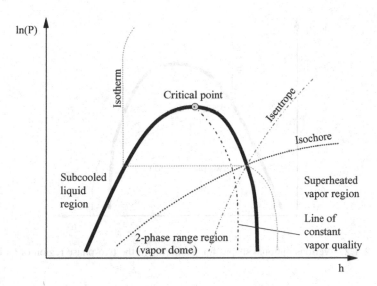

FIGURE 2.1.25 Schematic of ln(*P*)-*h* diagram.

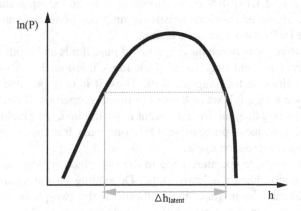

FIGURE 2.1.26 Latent heat (Δh_{latent}) in ln(*P*)-*h* diagram.

The ln(*P*)-*h* diagram is highly regarded in the refrigeration industry because it shows very quickly and clearly the amounts of heat exchanged or work required for the various processes. As long as all processes are steady state or steady flow processes, the change in enthalpy describes the amount of energy added or extracted. For example, the latent heat of a fluid for a given saturation pressure is determined by the enthalpy difference between saturated vapor and saturated liquid. In Figure 2.1.26, the latent heat is indicated. An inspection of Figure 2.1.26 shows that the latent heat decreases as the pressure increases and reaches the critical pressure. It should be noted that any difference in enthalpy is related to the addition or extraction of energy from the fluid, even if the process is not a constant pressure process. This is discussed in more detail in Section 2.2.

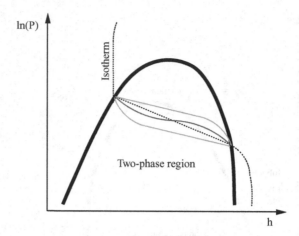

FIGURE 2.1.27 Various optional shapes for isotherms in the two-phase region of an $\ln(P)$-h diagram.

At first glance, the pressure-enthalpy diagrams for zeotropic mixtures have the same general topology as those for pure fluids. However, there are two important differences (Figure 2.1.18). R407C is chosen here as an example although it is a ternary mixture. However, the characteristics of multicomponent mixtures and binary mixtures in the $\ln(P)$-h diagram are the same.

The first difference between the diagrams of pure fluids and mixtures is that the zeotropic diagram is valid only for a given mass fraction. For a different mass fraction, all the lines on this diagram shift. Thus, it is only possible to show one diagram at a time with a full set of lines of constant properties. If one attempted to superimpose a second diagram for a different mass fraction, the graphical representation would become incomprehensible. Only one mass fraction is typically circulating in a vapor-compression cycle.

Second, due to the temperature glide in the two-phase region, the isotherms in the two-phase region have a negative slope. Depending on the nature of the pure fluids that constitute the mixture, the isotherm in the two-phase region may be concave, convex, s-shaped or a straight line. The schematic in Figure 2.1.27 shows examples of these isotherms. When one considers a constant pressure evaporation process that commences with saturated liquid and ends with saturated vapor, it becomes apparent that the beginning of the evaporation process starts at a given isotherm and ends at an isotherm of higher temperature. Thus, the temperature glide is correctly represented.

2.1.4 TEMPERATURE-ENTROPY DIAGRAM

Temperature-entropy (T-s) diagrams are traditionally used in the design of power generation plants. The only example for such a diagram that is frequently found in thermodynamics texts is the one for water and steam. Here the T-s diagram is shown primarily to indicate the differences between the one for pure fluids and the one for mixtures. Figure 2.1.28 shows the schematic of a diagram for a pure fluid. The two-phase

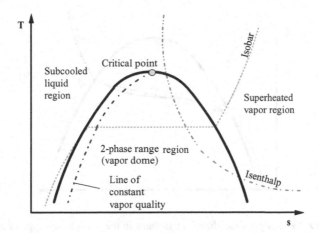

FIGURE 2.1.28 *T-s* diagram for a pure fluid.

range (vapor dome) has a similar shape compared to the mixture diagram. Again, the subcooled liquid range is located to the lower entropy (left) side of the vapor dome, the superheated vapor region is to the right of the vapor dome and the lines constituting the vapor dome are the saturated liquid and saturated vapor lines. They culminate in the critical point on top of the vapor dome. The isobar is horizontal in the two-phase region and ascends from the saturated vapor line with increasing slope into the super-heated vapor region. From the saturated liquid line reaching into the subcooled liquid region, the isobar drops very quickly and hugs the saturated liquid line, following it much more closely than shown in Figure 2.1.28, especially when pressures are low.

In a *T-s* diagram for a zeotropic mixture, all lines are similar to the pure fluid with the exception of the isobars. To correctly represent the temperature glide, the isobars must show a positive slope in the two-phase region. Depending on the nature of the pure fluids that constitute the mixture, the isotherm in the two-phase region may be concave, convex, *s*-shaped or a straight line. The schematic in Figure 2.1.29 shows examples of these isotherms.

When considering a constant pressure evaporation process that commences with saturated liquid and ends with saturated vapor, it becomes apparent that the beginning of the evaporation process starts at a given temperature and ends at a higher tem-perature on the same isobar.

2.1.5 TEMPERATURE-ENTHALPY DIAGRAM

Granyrd et al. (1991) suggested using temperature-enthalpy (*T-h*) diagram for zeo-tropic mixtures that have a temperature glide during their phase change processes. Benefits of *T-h* diagrams include ease of finding cycle problems, such as heat exchange pinch-points and wet-compression.

Figure 2.1.30 shows *T-h* diagrams of pure R22 and a zeotropic mixture R22/R114 (50/50 wt.%). Figure 2.1.30 shows different characteristics of isobaric phase change processes. The isobaric phase change lines have a zero slope for pure R22, but have nonzero slope for the zeotropic mixture R22/R114 (50/50 wt.%). Nonzero slope

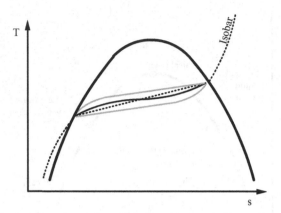

FIGURE 2.1.29 Various possible shapes of isobars in the two-phase range of a *T-s* diagram for zeotropic mixtures.

means a finite specific heat in the two-phase range. Moreover, the slope varies with respect to enthalpy; this means a variable specific heat resulting from a varying latent heat of vaporization due to the changing liquid and vapor mass fractions of constituents. The superheated region forms a narrow band as compared to other diagrams.

When the refrigeration cycle is superimposed together with temperature profiles of secondary fluid heat source and sink on the *T-h* diagram, the benefit of zeotropic mixtures is demonstrated graphically. Figure 2.1.31 shows the *T-h* diagram of a cycle with heat exchange operating 400 kPa evaporation and 1600 kPa condensing pressures for pure R22 and the zeotropic mixture R22/R114 (50/50 wt.%). In Figure 2.1.31, the cycle ABCD represents the refrigeration cycle (A–B: evaporation, B–C: compression, C–D: condensation, D–A: isenthalpic expansion) and the lines A_e–B_e and C_c–D_c represent temperature profiles of heat source and heat sink. When the temperature difference between the refrigerant and secondary fluid are compared for two refrigerants, as shown in Figure 2.1.31, advantages of zeotropic mixtures become distinct. For R22, the saturated vapor temperature during its condensing process is closest to the secondary fluid temperature, showing a potential heat exchange pinch-point. Moreover, R22 shows nonuniform approach temperatures during its evaporation process, whereas R22/R114 shows almost uniform and smaller approach temperatures than those of R22 during its evaporation and condensation processes. Therefore, approach temperatures at the inlet and outlet of both R22 heat exchangers are larger than those of R22/R114. Large approach temperatures mean higher irreversible losses associated with heat exchange. Therefore zeotropic mixtures having temperature glides that are properly matched with the secondary fluids have less irreversible losses than those of pure refrigerants.

2.1.6 Pressure-Mass Fraction Diagram

It should be noted that some designers prefer the pressure-mass fraction (*P-x*) diagram to the temperature-mass fraction diagram that was discussed in detail at the

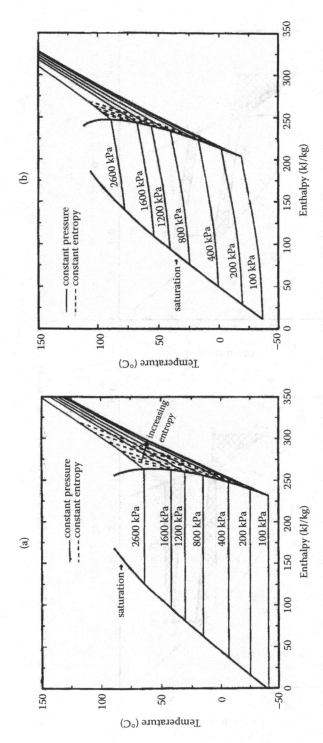

FIGURE 2.1.30 Temperature-enthalpy diagram (Granyrd et al., 1991). (a) R22, (b) R22/R114 (50/50 wt.%).

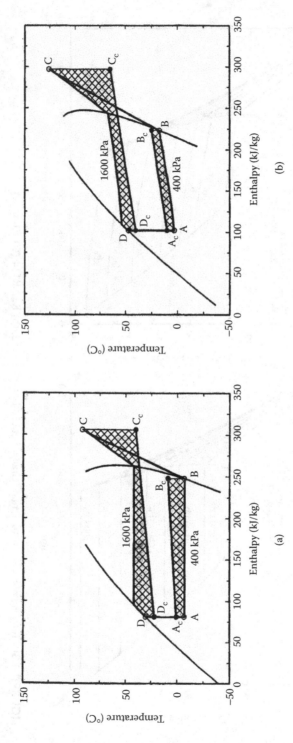

FIGURE 2.1.31 Temperature-enthalpy diagram of cycle with heat exchange (Granyrd et al., 1991). (a) R22, (b) R22/R114 (50/50 wt. %)

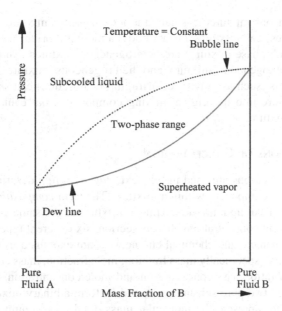

FIGURE 2.1.32 Pressure-mass fraction diagram.

beginning of this chapter. In principle, both diagrams are very similar except for the important difference that the general outline is reversed.

Figure 2.1.32 shows a schematic of a P-x diagram. The subcooled liquid region is now located above the two-phase region and the superheated vapor region is below. Thus, all processes shown in the diagram are reversed compared to the T-x diagram. Because this reversal can be quite confusing, it is recommended that readers pick either the temperature-mass fraction diagram or the pressure-mass fraction diagram as their favorite representation of mixture properties and then remain consistent. In this text, the temperature-mass fraction diagram is used exclusively.

2.2 ANALYTICAL TREATMENT OF THERMODYNAMIC PROPERTIES

This chapter focuses on the thermodynamic relationship between variables and their mathematical formulation.

2.2.1 TYPES OF SUBSTANCES IN INTERNAL EQUILIBRIUM

Substances in internal equilibrium can be divided into homogeneous and heterogeneous categories. A single-component substance is homogeneous when thermophysical properties are uniform for any arbitrary part of the substance. For single-component substances, single-phase substances are homogeneous and multiphase substances are heterogeneous. Combining pure substances forms a *mixture*. A mixture is homogeneous when the composition and thermophysical properties are uniform for any arbitrary small element of the mixture. When a mixture does not satisfy the conditions for the homogeneous mixture, it is called a heterogeneous mixture. For multicomponent substances, gas mixtures in thermodynamic equilibrium are homogeneous mixtures and multiphase mixtures are heterogeneous mixtures. Another difference between the homogeneous mixture and heterogeneous mixture is that simple mechanical means (such as gravity, centrifuging or filtration) can separate the heterogeneous mixture into its original mixing components, but cannot separate the homogeneous mixture.

2.2.2 DEFINITIONS OF CONCENTRATION

The following definitions are used in this text: *concentration* describes the relative amount of a given component within a mixture. The term *composition* refers to the components that make up a mixture. Thus, a mixture of the same composition can have many different concentrations. In this section, six different types of concentration used in mechanical and chemical engineering are explained using binary mixtures as an example, specifically mass fraction, mole fraction, mass component ratio, mole component ratio, mass concentration and mole concentration.

Let us assume two pure substances, A and B, form a binary mixture. We define M, m, m_i and V to represent the molecular mass, total mass, component mass and total volume, respectively. The six different types of concentration are defined as shown in Table 2.1. These concentrations are related to each other, so if one type of concentration is given, the other types of concentration can be calculated according to Table 2.2. In this text the mass fraction is used predominantly to describe a mixture concentration.

2.2.3 SOLUBILITY

Solubility is defined as the amount of a substance that dissolves in a specific amount of the dissolving agent. Two other terms that can describe the solubility are miscible and immiscible. Two gases form a homogeneous mixture when they are mixed. Similarly, many liquids, when suitable, are mixed and form a homogeneous mixture; these fluid pairs are *miscible*. Liquids that form an inhomogeneous mixture and do not dissolve are termed *immiscible*. Substances with the same polarity tend to be miscible and substances with opposite polarity tend to be immiscible. Good examples of miscible and immiscible refrigerant mixtures and lubricant pairs are shown in Table 2.3.

The miscibility depends on the substances that are mixed, the mass fraction and temperature. Figure 2.2.1 shows the miscibility of refrigerant R134a in a specific polyol-ester lubricant. The top of the curve point C indicates the minimum temperature at which the refrigerant and lubricant are miscible in all mass fractions. In the area to the left of the curve, the mixture is refrigerant rich and miscible, while in the area right of the curve, the mixture is lubricant rich and miscible. In the area below the curve the refrigerant and lubricant are present in two liquid phases. Each phase is a homogeneous mixture of refrigerant and oil but of different mass fractions. The area enclosed by the curve through A, B and C is termed the *mixing gap*. Their

TABLE 2.1
Types of Concentration for Binary Mixtures

Types of Concentration	Definition	Physical Meaning
Mass fraction (kg/kg)	Mass fraction of material B in the mixture $$x = \frac{m_B}{m_A + m_B} = \frac{m_B}{m}$$	1 kg of mixture contains x kg of material B and $(1-x)$ kg of material A
Mole fraction (mol/mol)	Mole fraction of material B in the mixture $$\Psi = \frac{m_B/M_B}{m_A/M_A + m_B/M_B}$$	1 mol of mixture contains ψ mol of material B and $(1-\psi)$ mol of material A
Mass ratio of components	Mass concentration of material B $$\beta = \frac{m_B}{m_A}$$	Amount of material B for 1 kg of material A
Mole ratio of components	Mole concentration of material B $$\chi = \frac{m_B/M_B}{m_A/M_A}$$	Molar quantity of material B in the $(1 + \chi)$ mol of the mixture as compared to that of material A
Mass concentration	Mass concentration of material B $$c = \frac{m_B}{V}$$	Amount of material B in 1 m³ of the mixture
Mole concentration	Mole concentration of material B $$c_m = \frac{m_B/M_B}{V}$$	Molar quantity of material B in 1 m³ of the mixture

Note: M, m and V stand for the molecular mass, mass and volume, respectively.

mass fractions of two liquid phases are given for a certain temperature by the horizontal limits of the curve for example points A and B at –60°C. Thus, at –60°C, R134a and polyol-ester lubricant are miscible when the lubricant mass fraction is either less than 4%, point A, or higher than 54%, point B. However, they are not miscible when the lubricant mass fraction is between the mass fractions of points A and B.

2.2.4 LIQUID MIXING PHENOMENA

When two liquids are mixed adiabatically, the mass is conserved but the volume and temperature are usually not. These phenomena are so-called mixing phenomena and can be explained with the help of Figures 2.2.2 and 2.2.3. When m_A kg of liquid A with a specific volume v_A and m_B kg of liquid B with a specific volume v_B form m_M kg of mixture, the mass, volume, enthalpy and entropy of the liquids before (index 1) and after (index 2) mixing are as follows:

TABLE 2.2
Transformation of Concentrations for Binary Mixtures

Given Concentration	Desired Concentration (see Table 2.1 for definitions)					
	x	Ψ	β	χ	c	c_m
$x = \dfrac{m_B}{m_A + m_B}$	x	$\dfrac{x}{\dfrac{M_B}{M_A}(1-x)+x}$	$\dfrac{x}{1-x}$	$\dfrac{M_A x}{M_B(1-x)}$	ρx	$\dfrac{\rho x}{M_B}$
$\Psi = \dfrac{m_B/M_B}{m_A/M_A + m_B/M_B}$	$\dfrac{\Psi}{\dfrac{M_A}{M_B}(1-\Psi)+\Psi}$	Ψ	$\dfrac{M_B \Psi}{M_A(1-\Psi)}$	$\dfrac{\Psi}{1-\Psi}$	$\dfrac{\rho \Psi}{\dfrac{M_A}{M_B}(1-\Psi)+\Psi}$	$\dfrac{\rho \Psi}{M_A(1-\Psi)+M_B\Psi}$
$\beta = \dfrac{m_B}{m_A}$	$\dfrac{\beta}{1+\beta}$	$\dfrac{\beta}{\dfrac{M_B}{M_A}+\beta}$	β	$\dfrac{M_A\beta}{M_B}$	$\dfrac{\rho\beta}{1+\beta}$	$\dfrac{\rho\beta/M_B}{1+\beta}$
$\chi = \dfrac{m_B/M_B}{m_A/M_A}$	$\dfrac{\chi}{\dfrac{M_A}{M_B}+\chi}$	$\dfrac{\chi}{1+\chi}$	$\dfrac{M_B\chi}{M_A}$	χ	$\dfrac{\rho\chi}{\dfrac{M_A}{M_B}+\chi}$	$\dfrac{\rho\chi/M_B}{\dfrac{M_A}{M_B}+\chi}$
$c = \dfrac{m_B}{V}$	υc	$\dfrac{M_A c}{\rho M_B + (M_A - M_B)c}$	$\dfrac{c}{\rho - c}$	$\dfrac{M_A\upsilon c}{M_B(1-\upsilon c)}$	c	$\dfrac{c}{M_B}$
$c_m = \dfrac{m_B/M_B}{V}$	$M_B V c_m$	$\dfrac{M_A c_m}{\rho + (M_A - M_B)c_m}$	$\dfrac{M_B c_m}{\rho - M_B c_m}$	$\dfrac{M_A c_m}{\rho - M_B c_m}$	$M_B c_m$	c_m

υ = specific volume of the mixture; ρ = density of the mixture.

Source: Bosnjakovic, 1965.

TABLE 2.3
Solubility of Refrigerant and Lubricant

Refrigerant	Lubricant	Solubility
R22 (nonpolar)	Mineral oil (nonpolar)	Miscible
R134a (polar)	Polyol-ester oil (polar)	Miscible
R134a (polar)	Mineral oil (nonpolar)	Immiscible

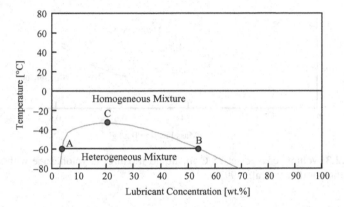

FIGURE 2.2.1 Miscibility of polyol-ester lubricant in R134a.

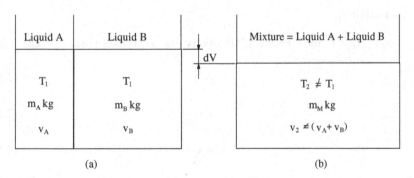

FIGURE 2.2.2 Liquid mixing phenomena. (a) Two liquids in one container, not mixed. (b) Liquids are mixed. The total volume is reduced in this example.

Before mixing:

$$m_1 = m_A + m_B = m_M \qquad (2.5)$$

$$v_1 = V_1/m_1 = (m_A v_A + m_B v_B)/m_M \qquad (2.6)$$

$$h_1 = (m_A h_A + m_B h_B)/m_M \qquad (2.7)$$

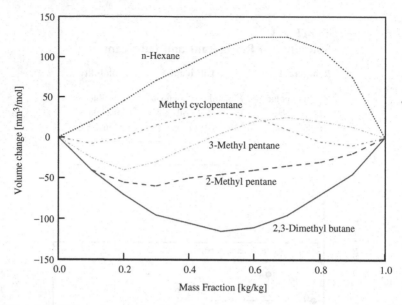

FIGURE 2.2.3 Volume changes at 25°C for liquid mixtures (Cyclohexane with some other C_6 hydrocarbons). (Perry et al., 1997.)

$$s_1 = (m_A s_A + m_B s_B)/m_M \qquad (2.8)$$

After mixing:

$$m_2 = m_1 = 1 \text{ (for the sake of simplicity, the total mass is set to 1 kg)} \quad (2.9)$$

$$v_2 \neq v_1 \qquad (2.10)$$

$$h_2 \neq h_1 \qquad (2.11)$$

$$s_2 \neq s_1 \qquad (2.12)$$

The volume change in the mixing of liquids can be measured easily and is usually small. Figure 2.2.3 shows the volume changes of liquid mixtures, cyclohexane with some other C_6 hydrocarbons, at 25°C. As shown in Figure 2.2.3, the changes of volume can be either positive or negative depending on the individual mixtures that are sensitive to the effects of molecular size and shape and to the differences in the nature and magnitude of intermolecular forces.

In general, the temperature of a mixture is different from the initial temperature of its components when the mixing process is adiabatic. During the mixing process, the mixture either rejects or absorbs energy called *heat of mixing*. When the heat of mixing is positive, the temperature of the mixture decreases. On the other hand, when the heat of mixing is negative, the temperature of the mixture increases. Figure 2.2.4 shows the heat of mixing for the H_2O/NH_3 mixture at 2 MPa and 40°C. The

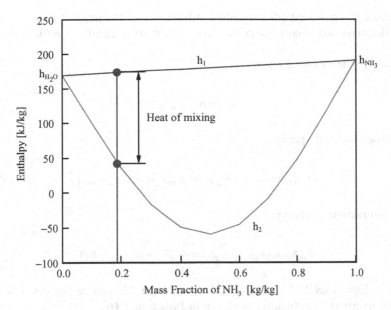

FIGURE 2.2.4 Heat of mixing in liquid mixture (H_2O/NH_3 mixture at 2 MPa, 40°C).

heat of mixing for the given mixture changes depending on the temperature and pressure. The heat of mixing for the refrigerants that are the focus of this text is quite small.

2.2.5 Adiabatic Mixing of Binary Mixtures

In this section, the adiabatic mixing process will be analyzed as shown in Figure 2.2.5. In Figure 2.2.5(a), two streams of binary mixtures, $\dot{m}_A$ of mixture A and $\dot{m}_B$

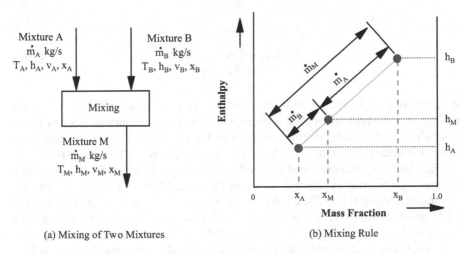

(a) Mixing of Two Mixtures   (b) Mixing Rule

FIGURE 2.2.5 Adiabatic mixing of two binary mixtures.

of mixture B, are mixed adiabatically and form $\dot{m}_M$ of the product mixture stream. From the mass and energy balances, three conservation equations result:

Conservation of total mass:

$$\dot{m}_M = \dot{m}_A + \dot{m}_B \tag{2.13}$$

Conservation of species:

$$\dot{m}_M x_M = \dot{m}_A x_A + \dot{m}_B x_B = \dot{m}_M x_A + \dot{m}_B (x_B - x_A) \tag{2.14}$$

Conservation of energy:

$$\dot{m}_M h_M = \dot{m}_A h_A + \dot{m}_B h_B = \dot{m}_M h_A + \dot{m}_B (h_B - h_A) \tag{2.15}$$

From Equations 2.13, 2.14 and 2.15, the ratio $\dot{m}_A / \dot{m}_B$ can be expressed in terms of mass fractions or enthalpies as shown in Equation 2.16,

$$\frac{\dot{m}_A}{\dot{m}_B} = \frac{x_B - x_M}{x_M - x_A} = \frac{h_B - h_M}{h_M - h_A} \tag{2.16}$$

The mass fraction (x_M) and enthalpy (h_M) of the mixture after mixing can be re-expressed as Equations 2.17 and 2.18 by using the ratio $\dot{m}_B / \dot{m}_M$,

$$x_M = x_A + \frac{\dot{m}_B}{\dot{m}_M}(x_B - x_A) \tag{2.17}$$

$$h_M = h_A + \frac{\dot{m}_B}{\dot{m}_M}(h_B - h_A) \tag{2.18}$$

As Equations 2.17 and 2.18 show, the mass fraction (x_M) and enthalpy (h_M) of the product mixture after mixing are a linear function of the two mixtures A and B and can be obtained by interpolation as shown in Figure 2.2.5(b). In an enthalpy-mass fraction diagram, Equations 2.17 and 2.18 describe the so-called *mixing rule*.

Similarly, the temperature of the product stream can be found from the enthalpy (h) and mass fraction (x) chart as shown in Figure 2.2.6. For example, two streams of saturated liquid R32/R134a, stream state A at $\dot{m} = 2/3$ kg/s, $x = 0.11$ kg/kg, $T = 30°$C, and stream state B at $\dot{m} = 1/3$ kg/s, $x = 0.75$ kg/kg, $T = 10°$C, are adiabatically mixed with each other. The state of mixture M can then be found from the mixing rule. The mass fraction and enthalpy of M can be calculated from Equations 2.14 and 2.15.

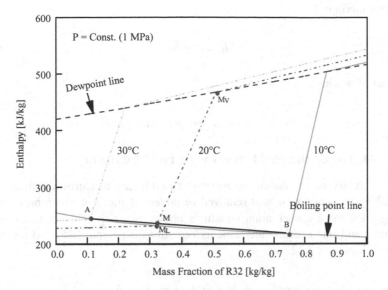

FIGURE 2.2.6 Enthalpy-mass fraction diagram of R32/R134a mixtures.

$$x_M = x_A + \frac{\dot{m}_B}{\dot{m}_M}(x_B - x_A) = 0.11 + 1/3(0.75 - 0.11) = 0.323 \ [\text{kg/kg}]$$

$$h_M = h_A + \frac{\dot{m}_B}{\dot{m}_M}(h_B - h_A) = 218.0 + 1/3(244.1 - 218.0) = 226.7 \ [\text{kJ/kg}]$$

The state of M can also be obtained from the $h\text{-}x$ diagram by interpolation as shown in Figure 2.2.5(b). In Figure 2.2.6, states A and B are connected by a straight line. Then the state point M of the product stream is found at the location where the line AB is divided in the ratio of AM:MB = 1:2. The temperature of the state point M is found from the isotherm TM that passes through point M. As can be seen in Figure 2.2.6, the state point M is above the saturated liquid line. Therefore, the state M is in the two-phase region.

2.2.6 ISOTHERMAL MIXING OF BINARY MIXTURES

When the temperature of the mixture is maintained before and after the mixing process, the process is called *isothermal mixing*. To be an isothermal mixing process, the mixture needs to be heated or cooled to maintain the initial temperature. The heat of mixing (Δh) for the isobaric and adiabatic process is obtained from the enthalpy difference between the enthalpies before (h_1) and after (h_2) mixing as follows:

Before mixing:

$$h_1 = (m_A h_A + m_B h_B)/m_M \tag{2.19}$$

After mixing:

$$h_2 = h_1 + \Delta h \tag{2.20}$$

Heat of mixing:

$$\Delta h = h_2 - h_1 = h_2 - (m_A h_A + m_B h_B)/m_M \tag{2.21}$$

2.2.7 Mixing of Binary Mixtures with Heat Exchange

If heat exchange occurs during the mixing of two binary mixtures, the final state is affected by the amount of heat removed or added. If the heat $\dot{q}$ is removed while mixing, then three conservation equations apply: two conservation equations for total mass and one species (the same as Equations 2.14 and 2.15) and the energy equation, Equation 2.22.

Conservation of energy: $\dot{m}_A h_A + \dot{m}_B h_B = \dot{m}_M h_M + \dot{q}$ \hfill (2.22)

Then the enthalpy of the mixture (h_M) becomes as Equation 2.23,

$$h_M = \frac{\dot{m}_A h_A + \dot{m}_B h_B - \dot{q}}{\dot{m}_M} = h_A + \frac{\dot{m}_B (h_B - h_A) - \dot{q}}{\dot{m}_M} \tag{2.23}$$

When the initial temperatures of the two mixtures are the same before mixing and maintained constant during the mixing process, the process is called *isothermal*. To be an isothermal mixing process, the mixture needs to be heated or cooled to maintain the initial temperature.

2.2.8 Specific Heat of Binary Mixtures

The enthalpy of a binary mixture can be defined using the heat of mixing (Δh) and the enthalpies of the individual components as follows:

$$h = \Delta h + \frac{\dot{m}_A h_A + \dot{m}_B h_B}{\dot{m}_M} = \Delta h + (1 - x)\, h_A + x h_B \tag{2.24}$$

The specific heat of the binary mixture can then be obtained by taking the derivative of enthalpy with respect to the temperature,

$$c_p = \left(\frac{\partial h}{\partial T}\right)_{P,x} = \left(\frac{\partial \Delta h}{\partial T}\right)_{P,x} + (1 - x)\left(\frac{\partial h_A}{\partial T}\right)_{P,x} + x\left(\frac{\partial h_B}{\partial T}\right)_{P,x} \tag{2.25}$$

From the definition of the specific heat of the individual components, Equation 2.25 becomes

$$c_p = \left(\frac{\partial \Delta h}{\partial T}\right)_{P,x} + (1-x)\, c_{p,A} + x c_{p,B} \qquad (2.26)$$

As indicated in Equation 2.26, the specific heat of binary mixture is obtained by summing up the mass fraction weighted specific heat of individual components and the change of the heat of mixing with temperature.

2.2.9 LATENT HEAT OF BINARY MIXTURES

The phase change of binary mixtures is an important engineering concern because the vaporization and liquefaction of binary mixtures are used in vapor compression heat pumps and are much different than those of pure substances. The *latent heat* for the phase change of a binary mixture can be defined as the enthalpy difference of saturated vapor (point M_V) and saturated liquid (point M_L) at constant pressure as shown in Figure 2.2.7. Figure 2.2.7 shows the enthalpy-mass fraction (*h-x*) diagram of the R32/R134a mixture at 1 MPa pressure. The differences between the saturated vapor enthalpy and saturated liquid enthalpy at the first and second *y*-axis represent the latent heat of evaporation of the pure R134a and R32, respectively. As an example, from the *h-x* diagram of R32/R134a, the saturated liquid and saturated vapor points (points M_L and M_V) at 0.5 kg/kg mass fraction of R32 can be found

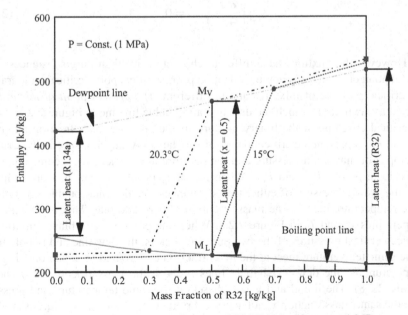

FIGURE 2.2.7 Enthalpy-mass fraction diagram of R32/R134a mixture at 1 MPa.

from the intersection of the vertical mass fraction line and the saturated liquid and vapor lines. The saturated liquid enthalpy is then 225.5 kJ/kg at 15°C and the saturated vapor enthalpy is 466.2 kJ/kg at 20.3°C. Therefore, the latent heat of the R32/R134a mixture at 1 MPa is 240.7 kJ/kg. In this way, the latent heat of the binary mixture at different pressures and concentrations can be obtained from the *h-x* diagram.

2.2.10　CRITICAL POINT OF BINARY MIXTURES

The *critical point* of pure substances is defined as the high pressure and temperature endpoint of the vapor pressure curve. Above the critical point (above the critical pressure and temperature), no liquefaction will take place. Moreover, the critical point is defined as a state where the saturation liquid and vapor states become the same as the pressure increases. The critical point is an inflection point where the curvature of the isobar in the *P-V* plane changes from positive to negative. The first and second derivatives of a curve at an inflection point are both zeros. Therefore, the first and second partial derivatives of the pressure with respect to volume at a constant temperature T_c are both zero at the critical point as shown in Equations 2.27 and 2.28,

$$\left(\frac{\partial P}{\partial V}\right)_{T=T_c} = 0 \tag{2.27}$$

$$\left(\frac{\partial^2 P}{\partial V^2}\right)_{T=T_c} = 0 \tag{2.28}$$

However, the mixtures have different characteristics than pure substances. The critical point of mixtures is not the inflection point. Critical point, critical temperature and critical pressure of mixtures are all different. The *critical point of mixtures* is defined as the point at which the dew line and bubble line meet. Figure 2.2.8 shows the critical envelope of the binary mixture in the pressure-mass fraction diagram. P_{CA} and P_{CB} represent the critical points of substances A and B, respectively. Critical envelopes of maximum pressure and temperature are formed while connecting the two critical points, P_{CA} and P_{CB}. When the pressure of a binary mixture is lower than the critical pressure of either two substances, the dew and bubble lines at the same temperature meet at the mass fractions of zero and one. This is the case for temperatures T_1 and T_2 in Figure 2.2.8. When the pressure of a binary mixture is between critical pressures of the two pure substances, the dew line and bubble lines at the same temperature meet at the mass fraction of zero and less than one. For the temperature T_4, the dew and bubble lines meet at the mass fraction of 0 and x_4 (Figure 2.2.8). The liquid and vapor phases at the same temperature and pressure have the same mass fraction x_4 at the critical point CP_4. If the temperature is slightly increased from point CP_4, while the mass fraction and pressure are maintained such that the temperature of the new critical point does not cross the critical envelope of

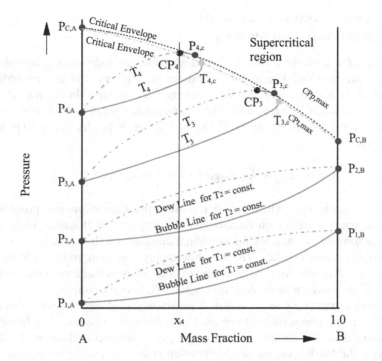

FIGURE 2.2.8 Critical envelope of binary mixture in pressure-mass fraction diagram (Bosnjakovic, 1965).

maximum pressure, the mixture is still in the two-phase region and not in the supercritical region. Therefore, the state of a mixture can be changed from the subcritical region to the supercritical region only when the pressure and temperature increase substantially such that the state point crosses the critical envelope of maximum pressure.

It should be noted that the condition where the second and third derivatives of molar Gibbs free energy with respect to the mixture composition are both zero is used in the REFPROP as the thermodynamic criteria for the critical loci of binary mixtures. Pseudocritical parameters are then fitted to empirical functions from the iteratively calculated critical loci (McLinden, 1998).

2.3 TERNARY AND MULTICOMPONENT MIXTURES

This text deals predominantly with binary mixtures (mixtures composed of two constituents). However, some of the preferred refrigerant mixtures under consideration are *ternary mixtures* (composed of three constituents). Table 2.4 shows a listing of these mixtures. The key thermodynamic characteristics that dominate the performance as refrigerants are not different from binary fluid mixtures. There is still a glide to contend with and that is the most prominent influence. However, the evaluation of thermodynamic properties and the representation of thermodynamic diagrams become considerably more complex.

2.3.1 THERMODYNAMIC DIAGRAMS FOR MULTICOMPONENT MIXTURES

The state of a pure fluid can be described with two independent variables, for example, pressure and temperature. To describe a binary mixture, an additional variable has to be introduced that describes the concentration. In this text, the mass fraction was selected. For a mixture with n constituents, we need $n-1$ mass fractions to fully describe all concentrations. The mass fraction of the ith component is defined as follows,

$$x_i = m_i / \Sigma m_j \text{ with } j \text{ ranging from 1 to } n, \text{ and } 1 \leq i \leq n$$

It is a considerable challenge to represent the thermodynamic properties of multicomponent mixtures on thermodynamic diagrams. If the mass fractions are chosen as some of the axes, multidimensional diagrams result that are hard to display on a two-dimensional surface. The most commonly used compromise is to show the conventional diagrams for pure fluids and to keep all mass fractions constant. Figure 2.1.18, the pressure-enthalpy diagram for R407C, is one example.

Ternary mixtures are an exception. A commonly used diagram exists that shows all three mass fractions on the three axes of a triangular diagram. The fourth axis, extending out of the surface, shows the property under consideration. Instead of displaying the fourth axis, its variable is often shown as a set of parametric curves superimposed on the surface of the triangle; Figure 2.3.1 displays examples of such a diagram. The corners of the triangle represent the constituents of the mixture. Each side of the triangle displays the binary mass fraction of the mixture composed of the constituents of the respective endpoints. Thus, each side has the same meaning as the x-axis used in the binary mixture diagrams in the earlier parts of Chapter 2, such as the temperature-mass fraction diagram. A point located on an axis of a triangular diagram describes the mass fraction of the respective binary mixture, as is the case for the point marked R410A in Figure 2.3.1(a). A point located somewhere within the triangle describes a ternary mixture, as with R407C in Figure 2.3.1(a).

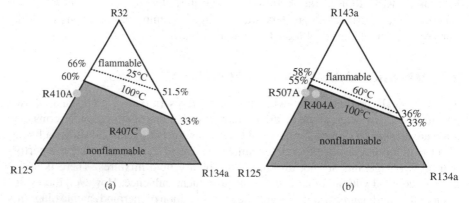

FIGURE 2.3.1 Flammability diagram for two ternary mixtures. (a) R407C. (b) R404A.

The application of ternary or multicomponent mixtures comes with the same challenges as that of binary mixtures. The major effects are caused by the temperature glide; thus, all the considerations in the following chapters hold accordingly. Only the mass balances must be extended to include additional equations to properly account for the additional components.

There are no known azeotropes among mixtures with more than two constituents.

2.3.2 FLAMMABILITY OF MIXTURES

The particular purpose of the triangular diagrams of Figure 2.3.1 is to describe the flammability of the mixtures as a function of mass fraction. In this case, the fourth axis is flammability and since the only information we are interested in is the isotherms that limit the regions in which the mixture is flammable or not are shown in Figure 2.3.1. Flammability depends on the temperature of the mixture (and other variables such as humidity not considered here) and therefore Figure 2.3.1(a) shows two isotherms for 25°C and 100°C and Figure 2.3.1(b) for 60°C and 100°C (UL, 1994).

Figure 2.3.1 also reveals that the fluids R32 and R143a are flammable and R125, R134a, R407C, R404A, R410A and R507A are not. Table 2.4 summarizes the properties of the pure fluids including flammability limits in air.

2.3.3 REASONING FOR SELECTING CERTAIN MIXTURES

There are a number of drivers that lead to the selection of certain mixtures. The overall intent is to develop an ideal refrigerant; that is, one that satisfies all the demands listed in Chapter 1, such as local and global safety (fluids should be non-toxic, nonflammable and have no adverse environmental impact) and should perform

TABLE 2.4
Thermophysical Properties of Pure Refrigerants

Property	R32	R125	R134a	R143a
Chemical formula	CH_2F_2	CHF_2CF_3	CH_2FCF_3	CF_3CH_3
Critical temperature [°C]	78.1	66.2	101.1	72.9
Latent heat[a] [kJ/kg]	299.0	124.7	190.7	177.7
Specific heat[a] [kJ/kg*K] (liquid/vapor)	1.806/1.367	1.310/0.958	1.370/0.946	1.551/1.164
LFL to HFL[b] [%]	12.7 ~ 33.5	None	None	7.1 ~ 16.1[c]
Molecular weight [kg/kmol]	52.0	120.0	102.0	84.0
Thermal conductivity[a] [W/m*K] (liquid/vapor)	0.147/0.0129	0.0663/0.0136	0.0876/0.0124	0.0765/0.0144

Note: NIST REFPROP V6.0 (McLinden et al., 1998) unless otherwise noted.

[a] at 10°C.
[b] Richard and Shankland (1992).
[c] Atofina Chemical (1999).

TABLE 2.5
Environmental and Thermophysical Properties of Binary and Ternary Mixtures

Property	R410A	R507A	R407C	R404A
Components	R32/125	R125/143a	R32/125/134a	R125/143a/134a
	(50/50 wt.%)	(50/50 wt.%)	(23/25/52 wt.%)	(44/52/4 wt.%)
ODP (R11 = 1.0)	0	0	0	0
GWP (CO_2 = 1.0)	1890[d]	HGWP 1.0[d]	1370[b]	3260[c]
Critical temperature [°C]	70.2	70.8	86.1	72.1
Latent heat[a] [kJ/kg]	208.4	152.2	199.2	156.0
Specific heat[a] [kJ/kg*K] (liquid/vapor)	1.565/1.207	1.427/1.057	1.449/1.009	1.434/1.057
Molecular weight [kg/kmol]	72.6	98.9	86.2	97.6
Thermal conductivity[a] [W/m*K] (liquid/vapor)	0.108/0.0129	0.0726/0.0141	0.0956/0.0121	0.0731/0.0140

Note: NIST REFPROP V6.0 (McLinden et al., 1998) unless otherwise noted.

[a] At 10°C.
[b] Calm (1995).
[c] Siva and Gustavo (1999).
[d] DuPont Fluorochemicals (2000).

well in heat pumps for all applications. Table 2.5 summarizes the environmental and thermophysical properties of the selected binary and ternary mixtures as most promising alternatives of R22 and R502. Moreover, constituents of these mixtures are two or three of R32, R125, R134a and R143a.

The properties of the new refrigerants should resemble as close as possible those of the fluids being replaced. In particular, for R22 no such candidate has been found. However, based on previous investigations that led to the introduction of R22, it was known that R32 is the most desirable fluid. It has the highest efficiency of the pure alternatives to R22 and, because of its high-pressure levels, also a very high volumetric capacity. Its only major shortcomings are its moderate flammability, the high compressor discharge temperature (a consequence of R32's low specific heat capacity) and the high-pressure levels. The latter can be seen as an asset or challenge, depending on the point of view.

The refrigerant mixtures R410A and R407C are compromises that aim at maintaining the favorable characteristics of R32, especially its high efficiency, while eliminating its shortcomings. R407C aims at achieving a fluid with properties resembling R22 as closely as possible, the only major difference being that R407C has a temperature glide. R410A, however, has almost no glide and it maintains the very high-pressure levels of R32.

R410A consists of R32/R125, 50/50 wt.%. R125 has almost identical characteristics to R32, especially the same pressure levels (Figure 2.1.8) but with a much higher specific heat capacity and it is not flammable. As a pure fluid, R125 performs quite poorly as refrigerant and the coefficient of performance (COP) is quite low.

This is a consequence of the very low critical point of R125, while that of R32 is rather high as compared to other refrigerants (Figure 2.1.8). The specific heat capacity of R125 is rather high and thus it contributes to a reduced compressor discharge gas temperature as compared to R32. Thus, R410A is nonflammable, has an acceptable compressor discharge temperature, a high volumetric capacity, good efficiency and a critical point that is lower than desirable. Under extreme air-conditioning situations, the critical point can be reached or exceeded.

R407C consists of R32/R125/R134a, 23/25/52 wt.%. With the addition of R134a, the vapor pressure is reduced and essentially matches that of R22. The mixture is not flammable, the compressor discharge temperature is considerably lower than that of R32 and the performance is acceptable. The critical point is higher than that of R410A and it is not prone to be reached under extreme air-conditioning applications.

R507A consists of R125/R143a, 50/50 wt.%. This mixture is designed as a long-term substitute for R502, a CFC-containing azeotrope. R404A consists of R125/R143a/R134a, 44/52/4 wt.%. With the addition of R134a, the suction vapor density matches that of R502. This mixture is designed to retrofit R502.

REFERENCES

Alefeld, G. and R. Radermacher, 1994, *Heat Conversion Systems*, CRC Press, Boca Raton, FL.

Allied Signal Chemicals, 1999, Genetron Products Brochure.

American Society of Heating, Refrigerating, and Air-Conditioning Engineers, Inc., 1997, *ASHRAE Handbook — Fundamentals*, pp. 19.8–19.60.

Atofina Chemical, 1999, Website: http://www.atofinachemicals.com/newelf/fluorochem/pdf/6049.pdf

Bosnjakovic, F., 1965, *Technical Thermodynamics*, Holt, Rinehart & Winston, New York, NY.

Calm, J.M., 1995, Refrigerant Database, ARTI Report, DOE/CE23810-59C.

Downing, R.C., 1988, *Fluorocarbon Refrigerants Handbook*, Prentice Hall, Englewood Cliffs, NJ.

DuPont Fluorochemicals, 2000, "Technical Information for Suva Refrigerants," Website: http://www.dupont.com/suva/na/usa/literature.

Fairchild, P.D., 2000, Personal communication.

Granyrd, E., J.C. Conklin and J.R. Sand, 1991, "Advantage of Enthalpy-Temperature Diagrams for Nonazeotropic Refrigerant Mixtures," Proceedings of XVIIIth Congress of Refrigeration, Montreal, Quebec, Canada.

Godwin, D.S., 1994, "Results of Soft-Optimized System Tests in ARI's R-22 Alternative Refrigerants Evaluation Program," Proceedings of The 1994 International Refrigeration Conference, West Lafayette, IN, pp. 7–12.

Herold, K.E., R. Radermacher and S.A. Klein, 1996, *Absorption Chillers and Heat Pumps*, CRC Press, Boca Raton, FL.

Hughes, M., 1997, "Contemporary Fluorocarbons, Refrigerant for the 21 Century," Proceedings of ASHRAE/NIST Refrigerants Conference, pp. 117–121.

Hwang, Y., J.F. Judge and R. Radermacher, 1995, "An Experimental Evaluation on Medium and High Pressure HFC Replacements for R-22," Proceedings of The 1995 International CFC and Halon Alternatives Conference, Washington D.C., pp. 41–48.

McLinden, M.O., S.A. Klein, E.W. Lemmon and A.P. Peskin, 1998, "REFPROP, NIST Thermodynamic Properties of Refrigerants and Refrigerant Mixtures Database," Version 6.0, Thermophysics Division of National Institutes of Standards and Technology, Gaithersburg, MD.

Perry, H.R., D.W. Green and J.O. Maloney, 1997, *Perry's Chemical Engineers' Handbook* (7th edition), McGraw-Hill, New York, NY, pp. 4–12.

Radermacher, R. and D.S. Jung, 1992, "Theoretical Analysis of Replacement Refrigerants for R-22 for Residential Uses," U.S. Environmental Protection Agency Report, U.S. EPA/400/1-91/041.

Richard, R. and L. Shankland, 1992, "Flammability of Alternative Refrigerants," *ASHRAE Journal*, Vol. 24, No. 4, pp. 20–24.

Sami, S.M., 1994, "A New Alternative HFC Blend for HCFC-22 and CFC-502," Proceedings of the 1994 International CFC and Halon Alternatives Conference, Washington D.C., pp. 56–64.

Siva, G. and D.R. Gustavo, 1999, "Total Equivalent Warming Impact of R22 Alternatives in Air-Conditioning and Heat Pump Application," *ASHRAE Transactions*, Vol. 105, Pt. 1, pp. 1228–1236.

Underwriters Laboratories Inc., 1994, UL Standard 2182, Refrigerants.

Vineyard, E.A., J.R. Sand and T.G. Statt, 1989, "Selection of Ozone-Safe, Nonazeotropic Refrigerant Mixture for Capacity Modulation in Residential Heat Pumps," *ASHRAE Transactions*, Vol. 95, Pt. 1, pp. 34–46.

3 Vapor Compression Cycle Fundamentals

The vapor compression cycle is the underlying thermodynamic process for most refrigeration systems in stationary and many mobile applications. This cycle is also referred to as the reverse Rankine cycle or the Perkins/Elmer cycle. For this text, the term vapor compression cycle is used throughout. As a general introduction and to establish the connection to the material taught in basic thermodynamics, the Carnot cycle will be explained first and then the vapor compression cycle will be discussed in full detail. This chapter also provides a description of the operation of heat pumps within the larger context of energy conversion systems.

3.1 THE CARNOT CYCLE

The *Carnot cycle* is an example of an idealized energy conversion cycle. Figure 3.1.1 shows a Carnot cycle for power generation on a temperature-entropy (*T-s*) diagram. Conventionally, in this chapter all temperatures are absolute values. The process line AB represents the isothermal addition of heat Q_2 at the temperature T_2 to a working fluid. We use here the convention that an arrow pointing to a process line represents energy supplied to the cycle. Line BC represents the isentropic production of work, CD the isothermal rejection of heat Q_1 at the temperature T_1 and DA the isentropic input of work. When all processes are reversible, the area enclosed by ABCD represents the net amount of work produced, W, and the area CDEF the amount of thermal energy Q_1 rejected by the cycle assuming E and F are located at T = 0 K. For a work-producing process, as shown in Figure 3.1.1, the processes follow a clockwise direction in the *T-s* diagram. The sum of the two areas CDEF and ABCD is the amount of heat Q_2 (thus area ABFE) supplied to the cycle as required by the First Law of Thermodynamics,

$$Q_2 = Q_1 + (W_{input} - W_{output}) \tag{3.1}$$

The signs of the quantities in Equation 3.1 are to be selected as indicated in Figure 3.1.1. All energies are counted as positive in the direction of the arrows. The *efficiency* for power generation is defined as the amount of work produced divided by the amount of heat supplied at the high temperature,

$$\eta = \frac{W}{Q_2} \tag{3.2}$$

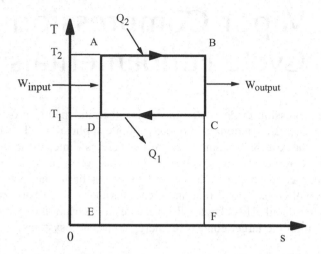

FIGURE 3.1.1 The Carnot cycle for power generation on a temperature-entropy diagram.

For the Carnot cycle of Figure 3.1.1, the Second Law of Thermodynamics states that for reversible operation the net entropy production is zero, so that

$$\frac{Q_2}{T_2} - \frac{Q_1}{T_1} = 0 \qquad (3.3)$$

Equation 3.2 can be modified to an expression that contains only temperatures using Equations 3.1 and 3.3 by eliminating W,

$$\eta = \frac{T_2 - T_1}{T_2} \qquad (3.4)$$

The expression in Equation 3.4 is frequently termed the *Carnot efficiency factor* for power generation.

Figure 3.1.2 shows a Carnot cycle that is operated as a heat pump cycle. The direction of all processes is reversed compared to the power generation cycle. Here, the temperatures are chosen such that heat Q_0 is added to the working fluid at T_0 along the process line GH, the fluid is compressed isentropically, HI, heat Q_1 is rejected at T_1, IJ, and the fluid is expanded isentropically, JG. Again, the areas represent energy transfers for reversible processes. The net amount of work input required for this cycle is represented by the area GHIJ and the amount of heat absorbed by the area GHKL. The sum of both areas (area IJLK) represents the amount of heat rejected at T_1.

The performance of a heat pump is described by the ratio of the benefit obtained — the amount of heat available at the high temperature — divided by the expenditure, the net work requirement. Since this value is always greater than 1.0, the term *coefficient of performance* (COP) is customary,

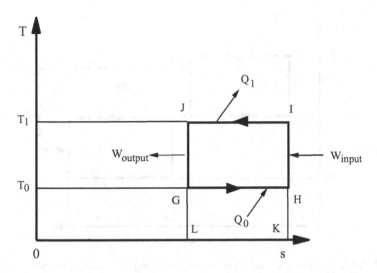

FIGURE 3.1.2 The Carnot cycle for heat pumping on a temperature-entropy diagram.

$$COP = \frac{T_1}{T_1 - T_0} \qquad (3.5)$$

This equation is derived by using the First and Second Laws for the heat pump process of Figure 3.1.2 and by following the same procedure as for Equation 3.4. The Carnot cycle of Figure 3.1.2 can also be used for cooling or refrigeration applications. While the cycle itself does not change, the way we view the application is different. For refrigeration, the heat removed at T_0 is of interest and the COP is defined as the ratio of the cooling capacity Q_0 over the work input. Subscript R is used to designate that the COP is defined for refrigeration,

$$COP_R = \frac{T_0}{T_1 - T_0} \qquad (3.6)$$

Equation 3.6 is derived by applying the First and Second Laws as described above for the power generation cycle. For the coefficients of performance of a heat pump cycle and refrigeration cycle, the following relationship can be derived. This expression can be verified by substituting Equations 3.5 and 3.6 into Equation 3.7,

$$COP_R + 1 = COP_H \qquad (3.7)$$

3.2 HEAT PUMPS IN THE CONTEXT OF ENERGY CONVERSION

To discuss the function of a heat pump within the larger context of energy conversion, a set of Carnot cycles is used. Now both cycles (the work producing and refrigeration

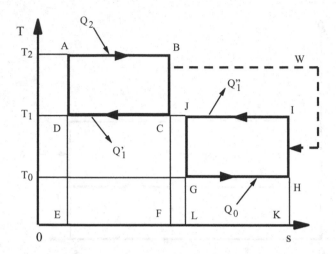

FIGURE 3.2.1 Carnot cycles for a combined power-generation/heat pumping facility such as an engine driven or absorption heat pump.

cycle of Figures 3.1.1 and 3.1.2) are considered together. It is assumed that the amount of work produced by the first cycle, Figure 3.1.1, is identical to the amount of work required by the second cycle, Figure 3.1.2. The new device, Figure 3.2.1, raises the temperature level of heat supplied at T_0 to T_1 by using the thermodynamic availability of the high temperature energy supplied at T_2. The waste heat of the power generation portion of this combined cycle is also rejected at T_1. Thus the total amount of heat rejected at T_1 has two contributions, Q_1' and Q_1''.

This combined cycle represents a heat pumping device that is driven by the input of heat only. This model is a reflection of the fact that most of the power available today is generated in a heat engine. Although this text focuses on the vapor compression cycle that is driven by the input of power, in a larger context primary energy made available mostly in the form of heat is the source of the power supplied to heat pumps. Thus, Figure 3.2.1 is the ideal representation of several different heat pumping concepts. Examples are an engine-driven vapor compression heat pump, a combination of a steam or gas turbine power plant with an electrical vapor compression heat pump or the representation of absorption or adsorption heat pump.

The coefficient of performance for the combined device is customarily defined as

$$COP_{AR} = \frac{Q_0}{Q_2} \tag{3.8}$$

for cooling and refrigeration applications and as

$$COP_{AHP} = \frac{Q_{1'} + Q_{1''}}{Q_2} \tag{3.9}$$

for heating applications.

By applying the First and Second Laws and eliminating either Q_0 or Q_1 (with $Q_1 = Q_1' + Q_1''$), Equations 3.8 and 3.9 can be converted into expressions that depend on temperature only:

$$COP_{AHP} = \frac{T_2 - T_0}{T_2} \frac{T_1}{T_1 - T_0}$$ (3.10)

$$COP_{AR} = \frac{T_2 - T_1}{T_2} \frac{T_0}{T_1 - T_0}$$ (3.11)

Here, the distinction between a heat pump and a refrigeration system is only a function of the application, not of the operating mode. Equation 3.7 is also valid for any heat driven heat pump:

$$COP_R + 1 = COP_{HP}$$ (3.12)

In Figure 3.2.1, it is implied that the power-producing cycle is operating at higher temperature levels than the heat pump cycle. This does not have to be the case. It is conceivable that applications call for the power being generated at temperature levels below those of the heat pump application. For example, a portion of waste heat from an industrial plant can be used to operate a power producing Rankine cycle and this power then drives a heat pump that lifts the remaining portion of the waste heat to a higher temperature level. Absorption heat transformers are an example of such systems.

3.3 THE IDEAL VAPOR COMPRESSION CYCLE

Figure 3.3.1 shows schematically the ideal vapor compression cycle. Figure 3.3.1(a) shows the cycle schematic and in Figure 3.3.1(b) the same state points are super-imposed on a $\ln(P)$-h diagram. Isentropic compression occurs from point 1 to 2 and isobaric heat rejection from point 2 to 3, representing the desuperheating of the vapor. The process from point 3 to 4 is the isobaric and isothermal heat rejection due to condensation and point 4 to 5a the isentropic expansion. The state points for the ideal vapor compression cycle can be located on the $\ln(P)$-h diagram with only minimal information about the operating conditions and the assumption of an ideal cycle.

The ideal vapor compression cycle used in this text is defined as follows: no pressure drop in any heat exchanger and connecting piping, saturated vapor is leaving the evaporator, saturated liquid is leaving the condenser, the expansion process is the isenthalpic process line 4–5 and the compression process is isentropic. When these conditions are fulfilled, one has to specify only two items of information to

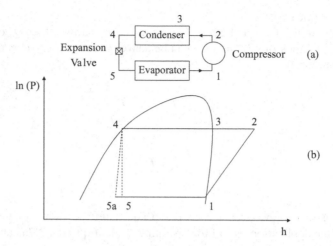

FIGURE 3.3.1 (a) Schematic of a vapor compression cycle. (b) Schematic of $\ln(P)$-h diagram with the state points of a vapor compression cycle superimposed.

plot the cycle on the $\ln(P)$-h diagram, the saturation pressure or temperature for the evaporator and the saturation temperature or pressure for the condenser.

To demonstrate this concept, the refrigerant ammonia is selected as an example. The saturation temperature of the evaporator is assumed to be $-20°C$ and that of the condenser $40°C$. The $\ln(P)$-h diagram is shown in Figure 3.3.2. With the knowledge of the evaporator saturation temperature, state point 1 is defined, located on the saturated vapor line at the intersection with the isotherm of $-20°C$. The saturation pressure can be read from the pressure axis as about 0.19 MPa. Points 3 and 4 are found from the knowledge of the condenser saturation temperature and the intersection with the saturated vapor and saturated liquid lines, respectively. The condenser saturation pressure is read to be 1.7 MPa. Point 2 is found by following a line of constant entropy from point 1 to the intersection with the isobar of the condenser saturation pressure. Since the expansion process is isenthalpic, point 5 is found vertically below 4, that is, the intersection of the line of constant enthalpy and the evaporator isobar. With this, the ideal vapor compression cycle is completely defined on the $\ln(P)$-h diagram. In most cases, it is best to start plotting the ideal vapor compression cycle on the $\ln(P)$-h diagram by finding points 1 and 4; then all others are fixed.

There seem to be two dominating reasons why the vapor compression cycle is the cycle of choice for a wide range of refrigeration applications. One reason is the fact that the latent heat during the liquid-vapor phase change process is used as the mechanism to transfer heat to and from the working fluid. This greatly reduces the mass flow rate in the cycle. But at least as important is the following consideration: All other refrigeration cycles involve the exchange of work twice, once in a compression process and once in an expansion process. The difference between these two contributions of work is the net work input for a heat pump. When each of the two processes that involve work has a realistic efficiency (for example, 0.8), the net work is to be divided by an overall efficiency of 0.64. This efficiency of the cycle

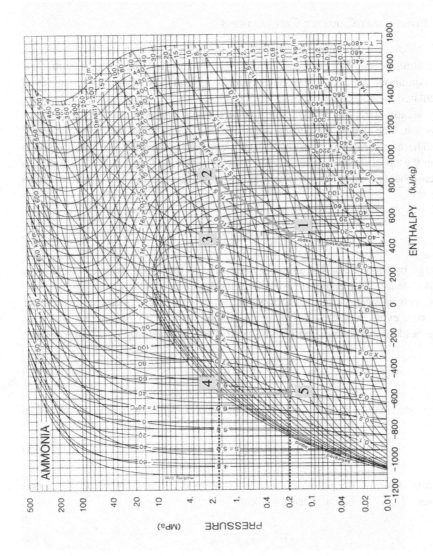

FIGURE 3.3.2 ln(*P*)-*h* diagram of ammonia vapor compression cycle. (Copyright 2001, American Society of Heating, Refrigerating, and Air-Conditioning Engineers, Inc. Reprinted by permission from ASHRAE 2001 Handbook — Fundamentals.)

is applied in addition to the thermal efficiency and represents significant performance degradation. The vapor compression cycle avoids this shortcoming with an elegant approach. One of the two isentropic processes that involve work, the expansion process, is conducted almost entirely in the liquid phase. Since liquids are essentially incompressible, only a minute amount of work is lost as compared to the expansion of a gas.

3.3.1 POWER AND CAPACITY CALCULATIONS

The $\ln(P)$-h diagram of Figure 3.3.2 is a very informative design tool that merits an understanding of more of its general features in order to quickly gain numerical information about the cycle. First, the amounts of heat and work that are involved are calculated.

Any steady-state thermodynamic analysis starts out with two basic laws of conservation. The first is the conservation of mass, the second the conservation of energy. The compressor is considered first: A mass balance for the compressor is based on the observation that for a steady state, steady flow process — assumed to be the case here — the mass flow that enters must also be leaving. Thus the mass balance states

$$\dot{m}_1 = \dot{m}_2 = \dot{m} \tag{3.13}$$

The energy balance states that all energy supplied to the compressor has to leave it also under steady state condition without considering compressor losses. The mass flow rate entering equals the mass flow rate leaving. In fact, since the ideal vapor compression cycle is a closed system as can be seen from Figure 3.3.1(a), this equation holds for all components of the cycle (W is replaced by Q_i as appropriate),

$$\dot{m}_1 h_1 + \dot{W} = \dot{m}_2 h_2 \tag{3.14}$$

Rearranging Equation 3.14 and using 3.13, the power input, W, can be calculated as

$$\dot{W} = \dot{m}(h_2 - h_1) \tag{3.15}$$

In a similar fashion, the amounts of heat exchanged in the evaporator, Q_{evap}, and condenser, Q_{cond}, can be calculated as follows:

$$\dot{Q}_{evap} = \dot{m}(h_1 - h_5) \tag{3.16}$$

and

$$\dot{Q}_{cond} = \dot{m}(h_2 - h_4) \tag{3.17}$$

The energy balance must also be fulfilled for the overall cycle. It is left to the reader to show that

$$\dot{Q}_{evap} + \dot{W} - \dot{Q}_{cond} = 0 \qquad (3.18)$$

Equation 3.18 serves as a good check to confirm that the above calculations are internally consistent.

Figure 3.3.2 reveals additional useful information. The temperature of the vapor leaving the compressor can be read from the chart, in this case about 141°C. Further, point 5 is located in the two-phase range. Using the lines of constant vapor quality, x, the quality at the evaporator inlet, is determined to be 0.22. This indicates that 22% of the stream entering the evaporator is already evaporated. This evaporation has to occur to ensure that the entire stream is cooled to the evaporator saturation temperature.

3.4 DIFFERENCES BETWEEN THE CARNOT CYCLE AND VAPOR COMPRESSION CYCLE

Figure 3.4.1 shows side-by-side the Carnot cycle (a) and the vapor compression cycle (b), each superimposed on a temperature-entropy diagram. Both are represented as ideal cycles for which realistic effects such as pressure drop or compressor inefficiencies are not accounted. The arrows indicate the direction in which the cycle progresses. There are two obvious ways that the vapor compression cycle deviates from the Carnot cycle. These are the so-called superheat horn and the expansion process. Both are shown in more detail in Figure 3.4.2. The superheat horn results from the fact that for most common refrigerants, the fluid leaving the compressor is superheated.

The *superheat horn* presents a performance penalty. It represents increased compressor work as compared to the Carnot cycle. This is indicated by the increased area enclosed by the cycle, which in a *T-s* diagram represents work. In this case of

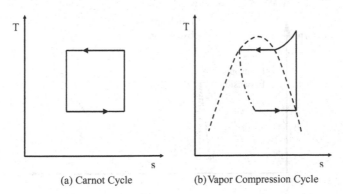

(a) Carnot Cycle   (b) Vapor Compression Cycle

FIGURE 3.4.1 Schematic of Carnot cycle (a) and vapor compression cycle (b) in a *T-s* diagram.

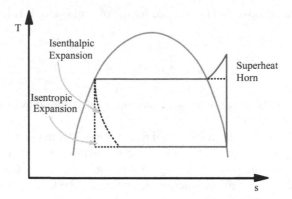

FIGURE 3.4.2 Isenthalpic and isentropic expansion for a vapor compression cycle in a *T-s* diagram.

a counterclockwise cycle, it represents compressor work input. The Carnot cycle is fluid independent and the superheat horn can be avoided. For the vapor compression cycle, it could also theoretically be avoided. For example, the evaporation process could be terminated at a quality selected such that the compressor outlet would be just saturated vapor. The compressor work input would be lower; however, the compressor suction would contain a considerable amount of liquid, a condition that would damage many compressors. The reason for the reduced compressor work is that in the case of the lower discharge temperature, less of the compressor work goes into heating the fluid and the compressor does not have to work as much against an increasingly hotter gas, as is the case when vapor alone is compressed. This observation also indicates that the specific heat capacity of the refrigerant vapor is an important variable influencing compression efficiency.

The expansion process for the vapor compression cycle is an isenthalpic process, while it is isentropic for the Carnot cycle. This simultaneously contributes two penalties, loss of work and loss of capacity. This is shown once more in Figure 3.4.3

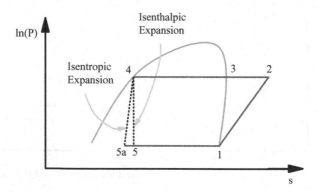

FIGURE 3.4.3 Isenthalpic and isentropic expansion for a vapor compression cycle in an ln(*P*)-*h* diagram.

in a $\ln(P)$-h diagram. The enthalpy difference $h_5 - h_{5a}$ represents the loss of work. This energy, when not extracted from the fluid, results in an increased vapor quality. Consequently, less liquid refrigerant is present to provide cooling.

It was shown in a previous chapter that the amount of work lost in the expansion process is small and usually readily accepted. The superheat horn is not easily avoided with the constraints imposed by conventional technology. Figure 3.4.4 puts these contributions more clearly in perspective (Domanski, 1995). It shows the relative penalties of the superheat horn and the expansion process. The term *relative penalty* is defined as follows: The relative penalty for the loss of cooling capacity is expressed as Q_{exp}/Q_C, which stands for "capacity with isenthalpic expansion" divided by the "capacity with isentropic expansion;" further, W_{exp}/W_C expresses the ratio of the respective work terms related to the expansion process and W_{sup}/W_C the additional work because of the superheat horn. In Figure 3.4.4, these losses are graphed on the same bar chart as stacked bars for a large number of refrigerants. The x-axis of Figure 3.4.4 shows the refrigerants sorted according to their vapor specific heat capacity. For ammonia, the refrigerant with the lowest heat capacity, the two work effects have more or less the same relative penalty. With increasing heat capacity, the effect of the superheat horn vanishes while both of the remaining penalties increase and the expansion process quickly becomes the most dominant one. It should be noted that the relative penalties are not indicative of how much each loss contributes to an overall degradation to COP. A more detailed analysis is given in Domanski (1995). Figure 3.4.4 shows how the penalties vary with the specific heat capacity of the refrigerant. This makes clear that the specific heat capacity of a refrigerant has an important influence on system performance, resulting in two effects. To better understand this, it is assumed for the following that a given refrigerant has a high specific heat capacity. It is assumed also, without limiting the general validity of the following consideration, that a positive displacement compressor is used.

The temperature increase during the compression process is low. This results from the fact that the compressor work is turned into the internal energy of the gas within the compressor. There are several options to increase the internal energy of a gas. Energy can be stored within molecules in the form of rotational or oscillatory motion or in the form of translational energy, i.e., the speed at which the molecule moves, its kinetic energy. Only the latter results in a temperature increase.

When the specific heat capacity of a gas is high, a considerable amount of the increase in internal energy is stored within the molecule and only a small fraction leads to an increase in kinetic energy, that is, its temperature. The temperature increase requires increased compression work. For fluids of high specific heat capacity, this effect is smaller than for fluids with low specific heat capacity. Thus, when the compressor compresses a gas that stays relatively cool, the compressor work is not as high as if the specific heat capacity were low and the temperature of the gas increased more quickly.

On the other hand, during the expansion process of a fluid with a large specific heat capacity, a relatively large amount of refrigerant has to evaporate to cool the remaining liquid to the evaporator saturation temperature and less liquid is left to provide cooling capacity.

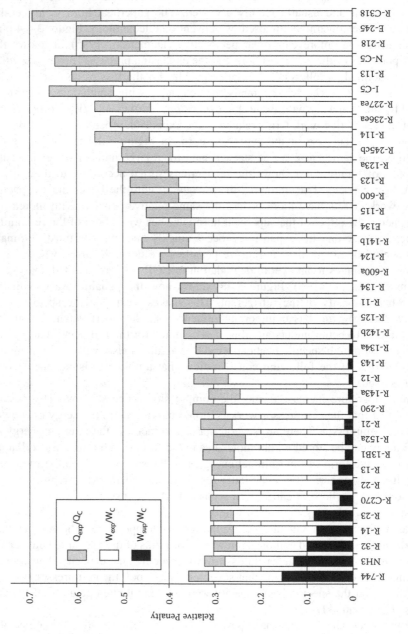

FIGURE 3.4.4 Relative penalties of the superheat horn and expansion process (Domanski, 1995).

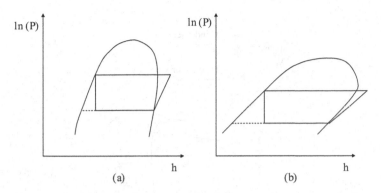

FIGURE 3.4.5 Schematic of vapor compression cycle in a ln(P)-h diagram for a fluid of low (a) and high (b) specific heat capacity.

This situation is better understood when the ln(P)-h diagram for a refrigerant of high specific heat capacity is compared to one with low specific heat. Figure 3.4.5 shows the two diagrams side by side. Figure 3.4.5(a) shows the diagram of a low heat capacity refrigerant; Figure 3.4.5(b), that of a high heat capacity fluid. Obviously, the vapor dome is tilted much more to higher enthalpies for the high capacity fluid. As a consequence, the expansion process produces much more vapor and less liquid and the evaporator capacity is reduced considerably. Furthermore, the compression stays very close to the saturated vapor line and the compressed vapor is almost not superheated at all. In fact, for some refrigerants, the state point of the compressed vapor might be located in the two-phase range. In practice, avoiding a two-phase stream leaving the compressor involves sufficiently superheating the suction vapor. The penalty in terms of increased compressor work is small or even desirable to avoid a two-phase condition at the compressor outlet.

3.5 REALISTIC VAPOR COMPRESSION CYCLES

In Section 3.3, the ideal vapor compression cycle was introduced. However, under realistic operating conditions, several effects must be considered that significantly impact the cycle performance.

A process scheme of a realistic vapor compression system is shown in a ln(P)-h diagram in Figure 3.5.1. This figure also shows a schematic of the vapor compression cycle. The state points in the schematic correspond to those in the ln(P)-h diagram. In comparison to the ln(P)-h diagram of Figure 3.3.1, the first major deviation that is immediately obvious is the slope in the process lines that used to be isobars. Since the refrigerant has to flow through the heat exchangers and any connecting piping, a certain pressure drop is necessary. This is reflected in the diagram of Figure 3.5.1 by the slope in the process line for the evaporation process, line 5–1, for the desuperheating process, line 2–3, and the condensation process, line 3–4a.

Starting with the evaporator outlet and following the refrigerant around the cycle, note that the refrigerant is superheated, line 1–1a. The superheating is desirable for two reasons: First, it ensures that all the refrigerant is evaporated where it can

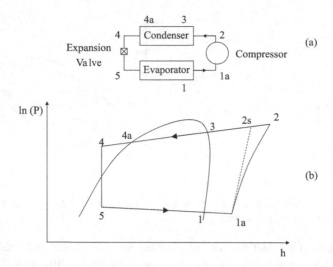

FIGURE 3.5.1 (a) Schematic of vapor compression cycle with state points to explain realistic operating conditions. (b) Schematic of realistic vapor compression cycle on a ln(P)-h diagram. Superheat (1–1a), subcooling (4a–4) and pressure drop (slant in line 2–4 and 5–1a) are included.

contribute to the cooling capacity. Second, superheating usually occurs within the evaporator for the purpose of ensuring that the compressor receives only vapor without any liquid entrained that could damage the compressor.

Due to the pressure drop in the evaporator and the superheating at the evaporator outlet, the compression process starts at a lower pressure and further into the superheated vapor region than in Figure 3.3.1. This is a double penalty for the compressor. A higher-pressure difference needs to be overcome by the compressor and the work also increases due to the higher temperature of the suction vapor. An inspection of the ln(P)-h diagrams, Figures 2.1.12 through 2.1.24, reveals that the slope of the isentropes decreases as the enthalpy increases, i.e., as the state point moves further away from the two-phase region. Thus, for isentropic compression, the compressor work increases once again as the entering suctioned gas is super-heated more.

Furthermore, the compressor must overcome the pressure drop of the refrigerant in the discharge line and the condenser, yet another increase in compressor work compared to the ideal cycle. The entire compression process is represented by the process line 1a–2. Since the compression process is usually irreversible to some degree due to internal friction of the mechanical parts, heat transfer and valve pressure drop, point 2 lies to the right of the endpoint 2s that would be achieved with isentropic compression. An additional penalty is increased volume, reducing capacity for a given displacement.

The process line 2–3 represents the desuperheating process in the condenser and line 3–4a the condensation process. Both exhibit the necessary pressure drop. Line 4a–4 represents the subcooled region, where the condensate is cooled to a lower temperature.

In many applications, it is important that the refrigerant leaving the condenser is subcooled. This ensures that only liquid and no vapor bubbles enter the expansion device, allowing for better flow control and smaller devices. Also, if vapor bubbles were to enter the expansion device, there would be less liquid refrigerant available in the evaporator to produce cooling while the compressor still has to compress that portion of the vapor that did not contribute to the cooling capacity. This represents a loss of capacity and efficiency. Alternatively, too much subcooling indicates that a considerable portion of the condenser volume is filled by single-phase liquid. Thus, the area available for heat rejection from the condensing fluid is smaller than it could be. As a consequence, the saturation temperature is increased along with the saturation pressure. The compressor work increases as a result.

It can be stated that the degree of subcooling at the condenser outlet is primarily determined by the amount of refrigerant charged to a heat pump and the setting of the expansion device primarily determines the degree of superheat at the evaporator outlet.

3.6 LORENZ CYCLE

The Carnot cycle is based on the assumption that the cycle accepts and rejects heat at a constant temperature level. However, in many applications, heat is supplied by or rejected to a fluid which changes its temperature. Examples are the chilled water produced by a chiller or the air or cooling water to which an air conditioner rejects heat. Figure 3.6.1 presents the schematic of such heat exchange processes for a given refrigerant using a T-s diagram. Process line 1–2 represents the evaporating refrigerant and process line 3–4 the chilled water being cooled. The latter is superimposed on the T-s diagram of the refrigerant merely for indicating the water temperatures in comparison to those of the refrigerant. Line 5–6 represents the condensing refrigerant and line 7–8 the cooling water as it is being heated. At points 1 and 4 and at points 5 and 8, a so-called *pinch point* can be observed where the temperature

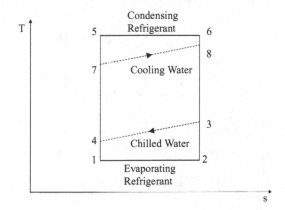

FIGURE 3.6.1 The evaporating temperature (1,2) and condensing temperature (5,6) are fixed by the lowest chilled water (4) and highest cooling water (8) temperature, respectively. A pinch point occurs at (1,4) and (5,8).

differences between both fluid streams become very small. As a consequence, the heat transfer rate becomes small as well. A pinch point limits the heat transfer rate due to very small approach temperatures. In order for a heat exchanger to still facilitate heat exchange, the available heat transfer surface area must increase. However, in this case the entropy production is also small due to the fact that heat is transferred over a small temperature difference. In contrast, at points 2 and 3 and also for 6 and 7, the temperature differences are large, the heat transfer rate is large as well and so is the entropy production.

Note that even when the heat transfer area approaches infinity, the performance improvement of the heat exchanger is limited (Rice, 1993a). While the temperature difference at the pinch point approaches zero, the temperature differences at the other end of the heat exchangers stay finite and the entropy production approaches a finite value. In case of the evaporation and condensing processes in Figure 3.6.1, the highest temperature of the evaporating refrigerant and the lowest temperature of the condensing refrigerant are fixed by the respective water temperatures. In turn, their pressures are fixed and thus the compressor power, so even with heat exchanger areas approaching infinity, the entropy production in the heat exchangers cannot be reduced to zero and the performance improvement of the entire heat pump is severely limited.

The *Lorenz cycle* addresses this issue (Herold, 1989; Alefeld, 1987). When a working fluid is available that changes its temperature during the course of the evaporation or condensing process (i.e., it has a temperature glide as described in Chapter 2), then the schematic of Figure 3.6.1 can be modified as shown in Figure 3.6.2.

In the ideal case that the change in temperature from the beginning to the end of the evaporation process matches that of the chilled water, both process curves become parallel and the pinch point is eliminated. Figure 3.6.2 shows the Lorenz cycle. With an increase in heat transfer area, the temperature difference between the two fluids continues to shrink everywhere in the exchanger and thus the overall entropy production can theoretically approach zero. Assuming that the cooling and chilled water temperatures are given, the suction and discharge pressures can

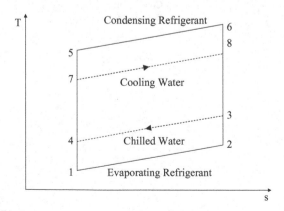

FIGURE 3.6.2 When the refrigerant saturation temperature has a glide, the pinch point can be eliminated.

approach each other to the largest possible extent, certainly much further than in the case shown in Figure 3.6.1. The overall efficiency of the cycle improves to the largest possible extent, too. It should be noted that even for the same heat exchange area, a Lorenz cycle that better matches the source and sink glides gives a higher cycle efficiency because it operates at an improved mean temperature than the corresponding pure refrigerant (Rice, 1993b).

The mixtures of Chapter 2 have the potential to approach the requirements of the Lorenz cycle depending on the degree to which they match the application glide in both the evaporator and the condenser. It is interesting to observe that for the given chilled and cooling water temperatures of Figure 3.6.2, the Carnot cycle, also symbolized by the process lines in Figure 3.6.1 (1–2–6–5), should have lower efficiency than the Lorenz cycle of Figure 3.6.2 (1–2–6–5). After all, in a *T-s* diagram, the area enclosed by the process represents the required work input as long as all processes are reversible. This condition is fulfilled here. The reader might have the expectation that no thermodynamic cycle can exceed the performance of the Carnot cycle.

Both statements are correct. Upon closer examination, the Lorenz cycle does not exceed the performance of Carnot cycles. This can be demonstrated with the help of Figure 3.6.3. The Lorenz cycle as shown in Figure 3.6.3 can be approximated by a series of Carnot cycles that are superimposed on the Lorenz cycle in a step-like fashion. In the limiting sense that the Carnot cycles have an infinitesimal length in the s-direction, they approach the Lorenz cycle.

Thus, the performance of the Lorenz cycle can be approximated through a series of Carnot cycles and therefore the Second Law is not violated. In general, the performance of many systems can be improved when a series of Carnot cycles (or more realistic cycles) are used that more closely match the temperature profiles of heat sinks and heat sources.

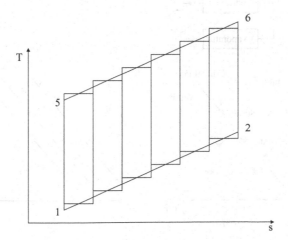

FIGURE 3.6.3 The Lorenz cycle (1–2–6–5) can be approached by many Carnot cycles (small, vertical rectangles), when operated such that they approximate the Lorenz cycle.

3.7 VAPOR COMPRESSION CYCLE WITH ZEOTROPIC MIXTURES IN THERMODYNAMIC DIAGRAMS

This chapter discusses the vapor compression cycle for zeotropic mixtures in light of the thermodynamic diagrams introduced in Chapter 2. This facilitates an understanding of the effects that mixtures have on the cycle performance. We focus on the ideal vapor compression cycle and binary mixtures for the sake of simplicity.

3.7.1 PRESSURE-TEMPERATURE DIAGRAM

Figure 3.7.1 shows the pressure-temperature diagram in two versions, for pure components (a) and for zeotropic mixtures (b). The mixture has the two constituents, A and B, and the vapor compression cycle is superimposed on each. The numbers for the state points are the same for both diagrams and for the inset of the vapor compression cycle schematic. For the pure fluid, all two-phase state points collapse onto the saturation line that separates the vapor and liquid regions. The state point 2 of the end of the compression process is located in the superheated vapor region. However, it becomes apparent that this diagram is very suitable for looking at overall operating conditions such as the relationships between pressure and temperature levels. It does not provide much detail about the processes within the heat exchangers.

For the mixture, we see more detail within the heat exchangers. Because of the glide, the inlet and outlet conditions are clearly distinguished and one can even see that the inlet quality to the evaporator has a value between 0.0 and 1.0. The meaning and location of state point 2, the compressor outlet, can be confusing and needs further explanation. Point 2 is plotted so that pressure and temperature of that state are correctly represented but it may be located within the solution field created by

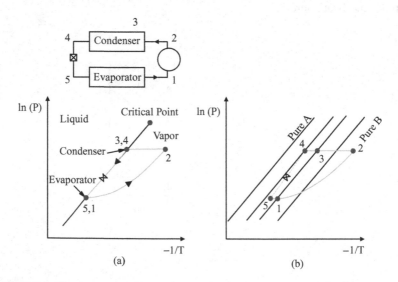

FIGURE 3.7.1 Vapor compression cycle superimposed on a $\ln(P)$-$(-1/T)$ diagram.

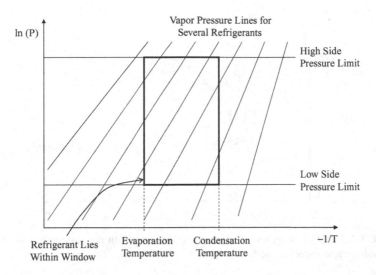

FIGURE 3.7.2 When a vapor pressure curve of a given refrigerant is located within the window, then it is suitable for the intended application. The window is specified through the desired evaporator and condenser temperatures and the acceptable high and low pressure limits.

the vapor pressure lines of the pure fluids A and B. This could imply that the outlet condition is a two-phase fluid. It must be understood that each of the vapor pressure lines is an isostere, a line of constant mass fraction. Each of the lines separates a liquid and a vapor region for a mixture of that particular mass fraction. Thus, when point 1 indicates saturated vapor leaving the evaporator, any point to the right of that particular isostere represents superheated vapor with regard to that isostere.

In most pressure-temperature diagrams, there are no provisions to indicate the dew points. The vapor curves shown indicate only bubble points. The user must find the information about dew points 1 and 3 from another diagram or source. Thus, the observer must understand the cycle and the diagram quite well to correctly comprehend the meaning of point 2.

The pressure-temperature diagram is nevertheless a good tool to gain insight about pressure and temperature relationships. For example, it allows a quick screening to determine which fluid is suitable for a given application. An application is determined by the expected temperature levels of the evaporator and condenser and by pressure limitations. The latter are dictated by two considerations: The first limitation is the maximum pressure rating for commercially available components and the second is given by what suction pressure is deemed acceptable for the compressor. These values create a window as shown in Figure 3.7.2.

3.7.2 PRESSURE-ENTHALPY DIAGRAM

Figure 3.7.3 shows the ln(P)-h diagram for a zeotropic mixture with superimposed state points of the vapor compression cycle. The topography of the cycle and its

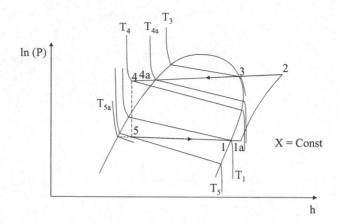

FIGURE 3.7.3 ln(P)-h diagram for a zeotropic mixture, with the state points of the vapor compression cycle superimposed.

state points are the same as those for pure fluids. The difference lies in the underlying fluid properties. Figure 3.7.3 shows, in addition to the cycle, the isotherms for the state points in the two-phase region. In this way the temperature glide is represented correctly. For example, in the evaporator the glide is T_1–T_5 and in the condenser T_3–T_{4a}.

Upon closer inspection, Figure 3.7.3 reveals other important differences between pure and mixed refrigerants in the vapor compression cycle. The temperature T_5 is higher than the liquid saturation temperature at the same pressure T_{5a}. This indicates that for mixtures, in contrast to pure fluids, the expansion process has two penalties. There is less liquid to evaporate, which is the case for pure and mixed fluids, but for mixtures, the temperature at the evaporator refrigerant inlet is not as low as it could be without the flashing. This first loss, caused by the increased vapor quality, affects the capacity and efficiency of the system and shows in the energy balance for the evaporator. It is therefore termed a "First Law loss." The second loss, the temperature increase, does not show up in the energy balance but its effects are clearly seen in a Second Law analysis. This situation is similar to the effects of the glide. The fact that refrigerant mixtures have a glide is not immediately visible in the equations of a First Law analysis using energy balances and is only apparent in a Second Law analysis of the evaporator or condenser.

Nevertheless, when there is a Second Law benefit such as less entropy production, it must show up in a better COP as well. With the glide, this benefit manifests itself by an increase in suction pressure or a decrease in the discharge pressure as discussed in Section 3.6; thus, the compressor work is reduced. The benefit that can be attained when the amount of vapor produced during flashing is reduced is discussed in more detail later.

The reader should keep in mind that if the overall mass fraction changes for any reason, this diagram is no longer valid and a new ln(P)-h diagram has to be generated for the new mixture mass fraction.

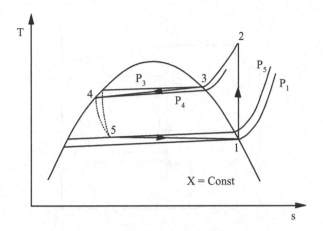

FIGURE 3.7.4 *T-s* diagram for vapor compression cycle. Pressure drop in the evaporator and condenser is indicated through the slanted process lines 5–1 and 3–4.

3.7.3 TEMPERATURE-ENTROPY DIAGRAM

Figure 3.7.4 shows the schematic of a *T-s* diagram for a mixture with the vapor compression cycle superimposed. In the two-phase region of the *T-s* diagram, the isobars have a positive slope. The line 5–1 shows clearly the pressure drop P_5–P_1 and how it offsets to some degree for the temperature glide, which would be more pronounced if the pressure drop were zero. For the evaporator, this pressure drop leads to a reduction in the temperature glide. While the temperature glide causes the temperature of the two-phase refrigerant to increase from inlet to outlet, the pressure drop causes this temperature to drop. These are two competing processes. Thus the temperature glide in the evaporator is reduced for two reasons: the flashing in the expansion valve and the pressure drop. These considerations influence the mixture selection for a given application or in turn the selection of the most appropriate temperature glide of the heat transfer fluid.

In the condenser, the situation is reversed. The temperature glide of the mixture and the glide caused by the pressure drop add up and the glide becomes larger. For these reasons and the difference in pressure, the glides in the evaporator are always smaller than those in the condenser. These effects are more obvious in this diagram than in the $\ln(P)$-*h* diagram described above.

In the *T-s* diagram, the influence of the compressor efficiency is more obvious. This is another example of how important it is to choose the most appropriate diagram to consider the question at hand. For a compressor of 100% efficiency, the compression process is isentropic and the process line 1–2 is vertical. Similarly, the expansion process is highly irreversible and the process line 4–5 reflects this as well. The entropy increases during the expansion process.

As for a $\ln(P)$-*h* diagram, the *T-s* diagram is only valid for a given mixture mass fraction. If this mass fraction changes for whatever reason, a new *T-s* diagram must be considered.

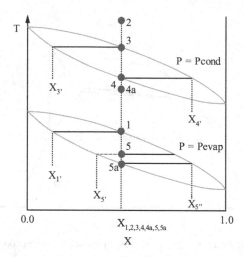

FIGURE 3.7.5 *T-x* diagram with the state points of the vapor compression cycle superimposed.

3.7.4 TEMPERATURE-MASS FRACTION DIAGRAM

Figure 3.7.5 shows a schematic *T-x* diagram of a binary zeotropic mixture with the vapor compression cycle superimposed. This diagram is rather unusual but it nevertheless reveals interesting features. The state points are again the same as in the previous diagrams. The mass fraction that circulates within the heat pump is x_1, which is the same for all state points, i.e., $x_1 = x_2 = x_3 = x_4 = x_{4a} = x_5 = x_{5a}$. Starting with point 1, the saturated vapor leaving the evaporator, one follows the refrigerant through the cycle. The compression process does not change the mass fraction but the pressure and temperature are increased and superheated vapor at point 2 is obtained. The vapor is cooled at constant mass fraction to saturated vapor, point 3, and the first liquid droplet appears with the mass fraction of x_3. At the end of the condensation process, point 4 is reached with a mass fraction of the last bubble of x_4. If the liquid is subcooled, it reaches point 4a. The expansion process ends at point 5, somewhere in the lower portion of the two-phase region at the evaporation pressure level. The refrigerant is in equilibrium with a liquid phase of a mass fraction x_5 and a vapor phase of mass fraction x_5. The evaporation process ends at point 1 with the last droplet having a mass fraction of x_1.

The *T-x* diagram is not well-suited to locating certain state points that are important for cycle design, such as points 2 and 5. But it demonstrates very well how the mass fractions change in the heat exchangers. This information is very important for detailed systems and heat exchanger design. The changes in mass fraction influence transport properties and therefore heat transfer coefficients and also the interaction with the compressor oil that is present in all parts of the system at a highly variable mass fraction.

Probably the most important feature of the *T-x* diagram is its ability to make intuitively clear that the mass fraction within the cycle can change for a number of reasons. Since during the evaporation and condensation processes the mass fractions of vapor and liquid are different, depending on its location, a system leak can cause a shift in the mass fraction of the remaining refrigerant mixture. The same holds for periods when the system is turned off. A heat pump is filled to about 30% of its total volume with liquid refrigerant. The remainder is refrigerant vapor with a mass fraction different from that of the liquid phase. Again, a leak that allows only vapor or only liquid to escape causes the mass fraction of the remaining refrigerant to shift. Since the liquid phase holds the bulk of the refrigerant mass, which is close to the charged mass fraction, vapor leaks are more effective in changing the mass fraction. The influence of leaks is discussed in more detail in Section 9.9.

3.8 THE MATCHING OF TEMPERATURE GLIDES

So far it is assumed that the temperature glides are linear. In Figure 3.6.1, straight lines indicate glides but, in general, glides are not linear. It should be noted that ternary mixtures tend to have a more linear glide than binary mixtures. Examples of nonlinear glides are indicated in the $\ln(P)$-h diagram schematic of Figure 2.1.27. This nonlinearity introduces interesting challenges for the design of heat exchangers.

   With a linear glide, the pinch point (the point of closest temperature approach between the hot and the cold fluid) must be located at either end of the heat exchanger or it does not exist when the two process lines are parallel. Possible cases are indicated in Figure 3.8.1 with one cold fluid stream and three different hot fluid streams. In general, there are three patterns of deviations, distinguished by their geometric characteristics: concave, converse and s-shaped temperature glides. For a given logarithmic mean temperature difference, the parallel case has the lowest entropy production. The location of the pinch point can be controlled by selecting the same thermal capacity (mass flow rate times specific heat capacity) for both fluids.

   Since the pinch point limits the heat transfer rate, its existence and location severely affect heat exchanger performance. The following consideration is limited to counterflow heat exchangers.

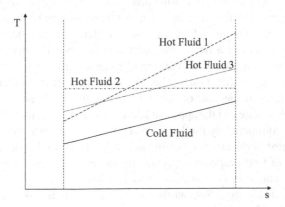

FIGURE 3.8.1 Matching of temperature glides. Hot fluid 3 has the best match with the cold fluid.

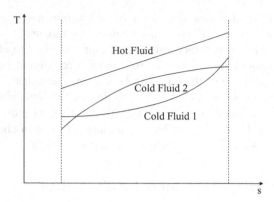

FIGURE 3.8.2 Matching temperature glides. Cold fluids 1 and 2 leave nonlinear temperature glides and pinch points can occur in the middle of the heat exchanger (Cold fluid 2) or in the end (Cold fluid 1).

When the glides are parallel, i.e., they "match" as it is often termed, the entropy production by the heat exchanger can be reduced to zero (in theory) by increasing the heat exchanger area. When a pinch point exists, this is not possible any more. When the heat exchanger area is increased, the approach temperature at the pinch point becomes smaller but a large temperature difference continues to exist at the other end of the heat exchanger. In the theoretical limit, for an infinitely large area, the entropy production of this heat exchanger stays finite, even when the approach temperature of the pinch point approaches zero. Thus, it is very important to strive wherever possible for matching glides in heat exchanger design. With a nonlinear glide, the pinch point can occur at any location within the heat exchanger. Figure 3.8.2 shows two examples with a concave and convex glide. For the concave case, the pinch point occurs at the end of the heat exchanger; however, for the convex case, the pinch point occurs somewhere in the middle. Its location can no longer be controlled by selecting the thermal capacity alone. Also, the existence of this pinch point cannot be observed when only inlet and outlet stream temperature are measured.

When refrigerant mixtures are used, a nonlinear glide can be thought of as caused by two mixture constituents of significantly different latent heats. Figure 3.8.3 displays the latent heat of a two-component mixture over the mass fraction range. The latent heat for component B is assumed to be significantly larger than that for component A. Let us further assume that the fugacity of B is smaller than that of A and that any effects of the heat of mixing are negligible. Thus, component A will predominantly evaporate and the apparent latent heat for this portion of the evaporation process will be dominated by the latent heat of A. This is indicated in Figure 3.8.3 with the latent heat of the mixture staying at levels similar to that of A. In this region, a small amount of heat evaporates a relatively large amount of refrigerant. With the accompanying change in composition, this accounts for a significant temperature glide as shown in Figure 3.8.4, while the remaining liquid is evaporated with a significantly higher latent heat.

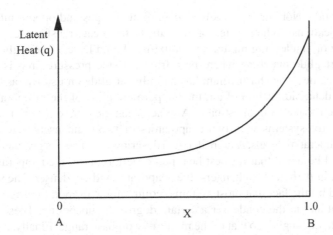

FIGURE 3.8.3 Component A has a low latent heat, component B a high one. When the mass fraction of A is high in the vapor phase, its latent heat dominated that of the mixture for a wide range of x.

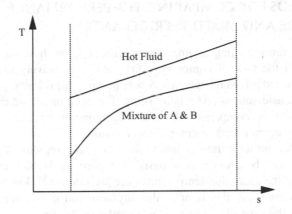

FIGURE 3.8.4 A mixture with the constituents A, B of Figure 3.8.3 shows a nonlinear glide with a pinch point in the middle of the heat exchanger.

During this first portion of the evaporating process, the remaining liquid phase will be enriched in B only when the evaporation process approaches its end and the remaining portion of B must evaporate as well with the latent heat of the mixture approaching that of B. The curve of the latent heat in Figure 3.8.3 rises to the level of the latent heat of B. In this portion of the evaporation process, the amount of heat absorbed for a given mass fraction change is larger. The resulting temperature interval is larger than before and the glide decreases in slope as shown in Figure 3.8.4. The mixture AB has characteristics that produce a convex temperature glide with a pinch point in the middle of the heat exchanger.

When the latent heat of A is much higher than that of B, the effects are reversed and curves with the opposite behavior were generated. The result is a mixture with a concave glide. Nonlinear glides have certain challenges and opportunities in heat pump applications. When water or dry air is the heat transfer medium for an evaporating or condensing mixture, the mixture should have a linear glide of equal size for best glide matching when the refrigerant side pressure drop is accounted for. When the heat transfer medium has a nonlinear glide on its own, as happens in the case of dehumidification of air, the temperature glide of the refrigerant mixture should match that of the air stream. Another example where a nonlinear glide is welcome is for systems with two evaporators or two condensers that operate on different temperature levels. A more detailed description of these is given in Chapter 5.

It should be noted that for best heat pump performance, the temperature glides must match in both heat exchangers, the evaporator and condenser. The situation is complicated by the fact that most systems require in the evaporator a certain degree of superheat and in the condenser a certain degree of subcooling. These character-istics add their own glide to that of the mixture two-phase range. Finally, a significant superheat is also present in the condenser. This challenge is never met by any heat transfer fluid and the designer is forced to find a suitable compromise.

3.9 METHODS FOR COMPARING THE PERFORMANCE OF PURE AND MIXED REFRIGERANTS

Since mixtures are emerging as important substitutes, the heat pump designer is confronted with the task of comparing fairly and conclusively the performance potential of pure and mixed refrigerants in heat pumps. The following chapter points out important considerations and relationships. The most important consideration is the basis on which the comparison is made. Depending on this choice, the outcome can vary greatly or even lead to contradictory results.

For example, the temperatures inside the two-phase regions of the condenser and evaporator can be chosen as a basis. For pure fluids, this choice may be acceptable since the saturation temperatures are well defined. However, when zeo-tropic mixtures are used, this is not valid anymore and the choice of the proper temperature for the sake of comparison is a matter of debate.

There are many different criteria that can be employed. From a user point of view, the heat pump with the lowest life cycle cost may be of interest but it could also be the one with the lowest operating or first cost. On the other hand, for an engineer conducting an exploratory study, a focus on the temperatures internal to the system may suffice for an initial estimate. Depending on the means and funds at hand, comparisons can span this range of complexity with varying results. The following is a list of basic assumptions on which comparisons are often based: life cycle cost of heat pump, first cost of heat pump, operating cost of heat pump, some heat exchanger area and some external fluid temperatures (i.e., chilled water) or refrigerant temperatures.

Alternatively, a user might not only focus on the life cycle or operating cost of a heat pump, but also on that of the complete system. For example, the first cost of making heat available, such as the earth loop for a ground coupled heat pump or the cooling water loop and cooling tower for a water chiller, is an important

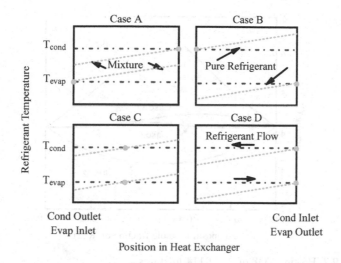

FIGURE 3.9.1 Four cases of refrigerant temperatures.

contribution. Most comparisons do not reach this level of sophistication because of the effort and expense involved. Major changes in common practice are difficult and expensive to implement.

The simplest criterion to base the comparison is to select the refrigerant temperatures in the evaporator and condenser. Because of the influence of the glide, this is where the confusion begins. Figure 3.9.1 shows four cases with the horizontal lines representing the two-phase temperature levels of the pure fluids, the dashed lines those of the mixture. An inspection of Figure 3.9.1 already reveals considerable differences. While for all four options the temperature lift is constant for pure fluids, it changes considerably for the mixture. Case A has the smallest average lift, while case B exhibits the largest. Cases C and D represent compromises that seem to produce average temperature lifts that are similar to those of the pure fluid. Since the temperature lift is in good first approximation a measure for the work requirement of the compressor, it is used here as a criterion to estimate the efficiency of the cycles. Thus, Case A greatly favors mixtures with a small lift and resulting high coefficient of performance, while Case B greatly disfavors mixtures. However, both cases are not realistic for most applications. Cases C and D represent a compromise but they can mask the performance potential of mixtures since they essentially force the average temperature lift to be equal for all cases. These initial observations are confirmed with more detailed calculations. McLinden and Radermacher (1987) employed a simple thermodynamic model using a mixture of R22 and R114 as an example.

Figure 3.9.2 shows the COP of pure R22, pure R114 and mixtures of both for various compositions. The cycle is specified by the various pairs of refrigerant temperatures of Figure 3.9.1. The saturation temperature of the pure components was selected as –10 and 50°C for the evaporator and condenser, respectively. Both pure components have approximately the same COP of 4.1; however, the COPs of the mixtures vary dramatically. For Case A, the mixture performs better than the

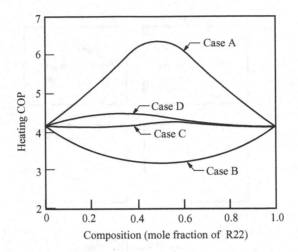

FIGURE 3.9.2 Heating COP of R22/R114 mixtures.

pure components, showing a maximum COP at a composition of about 50 mol% R114. For Case B, the COP of the mixture is always lower than that for the pure components. This result confirms the expectations stated above. Thus, Case A represents a hypothetical maximum improvement that can be achieved with mixtures as compared to pure refrigerants. Case B represents the highest performance degradation. COP comparisons based on the internal (refrigerant) temperatures are not conclusive and essentially worthless. However, the understanding of where the temperature glides should be for beneficial effects is instructive. In heat exchangers with counterflow heat exchangers and matched glides, the results are expected to be between Case A and Case C. In crossflow heat exchangers with poorly matched glides, the results will generally be between Case B and Case C.

It is a more productive approach to choose as the criterion for comparison the external temperatures, i.e., the temperatures of the heat transfer fluids. As long as these (and the flow rates) are constant, the heat pump performs the same duty and any change in the COP due to change in working fluids is clearly an efficiency gain or loss. However, when the external fluid temperatures (inlet and exit) are kept constant, the average temperature lift can still be influenced by the approach temperatures. To eliminate this uncertainty, the heat transfer area should be kept constant as well. This has the advantage that the cost for the heat exchangers is also essentially kept constant. As a consequence, this method approaches on first approximation a comparison at constant first cost. It is assumed here that the compressor cost and the heat transfer coefficients are the same for all fluids concerned; the latter is generally not the case. Nevertheless, on the basis of constant heat transfer area and coefficients and that the external temperatures are matched, the following observations can be made based on the assumption that the effects of superheating and subcooling, nonlinearity of the glide and influence of the expansion valve are neglected. It is further assumed that all heat exchangers are counterflow heat exchangers.

The results of this study can be summarized as follows:

1. When the temperature glide of the refrigerant inside the heat exchanger matches that of the external fluid stream, the potential COP improvement is largest.
2. If the glide of the refrigerant is much smaller or much larger than that of the external fluid, the COP is penalized.
3. The larger the glide (when refrigerant glide and external glide are matched) and the smaller the lift, the higher the improvement in COP.
4. The theoretical limit for the improvement in COP is the doubling of the value for a pure fluid (Case A in Figure 3.9.1). In this case, the temperature of the refrigerant leaving the evaporator is the same as that of the refrigerant leaving the condenser. Any further increase in the evaporator outlet temperature, or decrease in the condenser outlet temperature, would allow for direct heat exchange between the respective external fluid streams and eliminate the need for a heat pump in this range of the temperature glide.
5. Since the glide of a mixture varies with mass fraction, only certain mass fractions lead to the best match and therefore performance improvement. When a mixture has a large temperature glide and the application only a narrow one, there may be two mixture mass fractions that match the glide: a low mass fraction and a high mass fraction. At other mass fractions, the performance could be reduced.
6. The capacity changes primarily with the vapor pressure of the refrigerant at evaporator conditions. When mixtures have a large glide, the change in capacity with mass fraction can be considerable as discussed in Chapter 7.
7. The highest performance improvement is gained when the glide in the condenser and the evaporator are matched simultaneously.

3.10 SIMULATION OF THE VAPOR COMPRESSION CYCLE

This section introduces various modeling efforts for the vapor compression cycle. Modeling of the vapor compression cycle is important in determining the potential of alternative refrigerants and for designing the refrigerant cycle components and system appropriately according to the thermophysical characteristics of refrigerant mixtures since the experimental approach is too costly and time-consuming.

3.10.1 REFRIGERATION COMPONENT MODELS

Four components constitute a basic vapor compression cycle. Therefore, each basic cycle model consists of four component models: a compressor model, an expansion device model and two heat exchanger models. These component models serve as the basic building blocks for the cycle model.

3.10.1.1 Heat Exchanger Model

In order to develop a high quality cycle simulation for a vapor compression system, an accurate heat exchanger model is essential. The heat exchanger model significantly impacts not only the heat transferred, but also the amount of refrigerant in the system. In general there are two types of heat exchanger models: One is a relatively simple UA model and the other is a detailed model based on distributed parameter heat exchanger simulation that solves the continuity, species, momentum and energy equations simultaneously for air (or water) to refrigerant heat exchangers while using local heat transfer and pressure drop correlations.

Prominent among the previous work on this subject is the tube-by-tube analysis developed by Domanski (1989 and 1995). In this analysis, a wet (moisture from air conditioning on the outside surface) or dry evaporator was modeled by assuming the air passes directly from one tube to another. The heat and mass transfer of each tube was evaluated based on its own air and refrigerant inlet states; this allowed for investigation of air maldistribution. The simulation, restricted to pure refrigerants, was experimentally verified for wet coils and found to predict the overall capacity within −8.8 to +23.3% (Chawalowski et al., 1989).

Using the work by Domanski, Conde and Sutter (1992) developed a similar simulation with several modifications. One of the improvements included the ability to determine the refrigerant inventory within the evaporator. This code was part of an overall steady state cycle simulation and was experimentally verified with R22. The verification showed that the simulation frequently overpredicted the capacity by more than 10%.

Oskarsson et al. developed three steady state models for dry, wet and frosted evaporators. All of the models assumed that the entire heat exchanger could be simulated with one equivalent tube in crossflow. Furthermore, these models were developed for pure refrigerants. One of the models divided the evaporator into three different regions: the two-phase region, the transitional region and the superheated region. The most complex model divided the heat exchanger into finite elements, while the simplest model was parametric and required one of the other models to generate constants for it. The three-region model was experimentally verified and found to predict the capacity of a dry coil between −17.7 and +11.3%. Furthermore, the finite element model agreed with the three regions model within 2%.

Nyers has developed a transient evaporator model (Nyers, 1994). This model was developed for air to pure refrigerant applications. This model was fully distributed and simultaneously solved the continuity, momentum and energy equations but was not compared to any transient or steady state experimental data.

Judge and Radermacher (1997) developed a finite element transient and steady state model that was verified experimentally within 8%. The authors used this model to quantify the effects of heat exchange geometry on the performance of R407C.

For the conventional shell-and-tube heat exchangers, Hwang and Radermacher (1998) developed a water-to-refrigerant evaporator and condenser models that divided the heat exchanger into finite elements. These models are verified with experimental data for R22 and CO_2: a 5% deviation for the evaporator model and a 7% deviation for the condenser model were reported.

UA Model

When analyzing the heat exchanger, either *log-mean-temperature-difference* (LMTD) or the effectiveness-NTU (number of transfer units) method can be used.

In this section, only the LMTD model is explained. If the reader is interested in the effectiveness-NTU method, the detailed explanation and examples can be found from Incropera and DeWitt (1990).

In the UA model, the heat exchanger performance is defined as Equation 3.19,

$$Q = UA \times LMTD \tag{3.19}$$

where

U = overall heat transfer coefficient
A = total heat transfer area
$LMTD$ = log-mean-temperature-difference

In this model, three parameters in Equation 3.19 are given and the other variable is calculated by Equation 3.19. If the heat exchanger geometry (A) and the operational conditions (U and ΔT) are given, the heat exchanger performance (Q) is calculated. On the other hand, if the target performance (Q) is given, only one parameter among the heat exchanger geometry (A) and the operational conditions (U and ΔT) is calculated while the other two parameters are given.

Applying the UA model for each flow regime can further advance the UA model. An evaporator should be divided into two-phase and superheated regions. A condenser should be divided into superheated, two-phase and subcooled regions. The total heat exchanged is then the sum of the heat exchanged at each region, as shown in Equation 3.20,

$$Q = \sum Q_i = \sum U_i A_i \times LMTD_i \tag{3.20}$$

where

U_i = overall heat transfer coefficient for each segment
A_i = total heat transfer area for each segment
$LMTD_i$ = log-mean-temperature-difference for each segment

When applying the LMTD, some considerations must be accounted. First, the LMTD is designed for counterflow heat exchange configuration. However, it can be used for many different HX (heat exchanger) geometries by using the correction factor (CF) as shown in Equation 3.21,

$$LMTD = CF \times \frac{\Delta T_{in} - \Delta T_{out}}{\ln\left(\Delta T_{in} / \Delta T_{out}\right)} \tag{3.21}$$

where

CF = correction factor for heat exchanger geometry (examples are given in Figure 3.10.1)
ΔT_{in} = temperature difference between hot fluid inlet and cold fluid outlet $(T_{hf,in} - T_{cf,out})$
ΔT_{out} = temperature difference between hot fluid outlet and cold fluid inlet $(T_{hf,out} - T_{cf,in})$

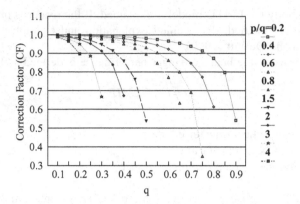

FIGURE 3.10.1 LMTD correction factor for single pass crossflow heat exchanger (one fluid mixed, other unmixed).

When one fluid undergoes phase change or two fluids have a very similar heat capacity, then $CF = 1$. A correction factor for single pass crossflow HX, one fluid mixed, other unmixed, as is the case for most of the air-to-refrigerant heat exchangers, is available from Bowman (1940) as shown in Equation 3.22,

$$CF = r/r_0 = r \left/ \frac{p-q}{\ln\left(\frac{1-q}{1-p}\right)} \right.$$

(3.22)

where $p = \dfrac{T_{hf,in} - T_{hf,out}}{T_{hf,in} - T_{cf,in}}$

$q = \dfrac{T_{cf,out} - T_{cf,in}}{T_{hf,in} - T_{cf,in}}$

Figure 3.10.1 shows the result of this correlation.

When applying the LMTD method for zeotropic mixtures having a nonlinear temperature glide during phase change, segmentation of the two-phase region is desirable to simulate the zeotropic mixture's phase-change correctly. When applying the LMTD method for multiphase region heat exchangers, the overall temperature difference can be defined in different ways. In Cycle 7, McLinden (1987) introduced the superposition LMTD as defined in Equation 3.23,

$$LMTD_{hx} = \sum_i f_i LMTD_i$$

(3.23)

where
    $LMTD_{hx}$ = LMTD for overall heat exchanger
        f_i = heat transfer fraction for individual phase zone (Q_i/Q_{hx})
    $LMTD_i$ = LMTD for individual phase zone

By assuming the constant heat transfer coefficient (U) for all phases, Rice and Sand (1990) and Domanski and McLinden (1990) introduced the harmonic weighted LMTD as defined in Equation 3.24,

$$\frac{1}{LMTD_{hx}} = \sum_i \frac{f_i}{LMTD_i} \qquad (3.24)$$

Detailed Heat Exchanger Models

Two heat exchanger models for the air-to-refrigerant (pure fluids and mixtures) heat exchangers, applicable to both pure fluids and mixtures, are introduced based on Judge and Radermacher's (1997) work: one for simulating the performance of the evaporator with two-phase boiling and single-phase heat transfer, one for simulating the performance of the condenser with two-phase condensation and single-phase heat transfer. It is assumed that the conventional finned tube heat exchangers are used. These models are distributed as parameter heat exchanger models. In this model, many existing heat transfer correlations were compared with the experimental data. As the boiling heat transfer correlations, Gungor and Winterton (1986), Shah (1982), Kandlikar (1991), Chen (1966) and Jung et al. (1989a) correlations were examined. As the condensation correlations, Tandon et al. (1986), Chen et al. (1987), Traviss et al. (1973), Shah (1979) and Dobson et al. (1994) correlations were examined. The authors recommended two correlations among the examined correlations: Jung's correlation predicted the evaporator capacity with −5 to +4% and Dobson's correlation predicted the condenser capacity within −3 to +7%. The increased deviation in condensation relative to evaporation occurred due to the lack of the condensation heat transfer correlations that were generally applicable for any mixture. However, the inaccuracy of the refrigerant side heat transfer coefficient is less significant than the airside. Details of boiling and condensation heat transfer correlations are introduced later in Chapter 8.

These heat exchanger models determine cooling or heating capacity by assuming two cycle parameters (evaporating pressure and condensing pressure) and calculating the pressure drop in the heat exchanger. Moreover, these models calculate required heat exchanger areas by calculating heat transfer coefficients and mean temperature difference between refrigerants and air. When the difference between the calculated heat exchanger area and the given heat exchanger area becomes less than the converging criteria, all cycle parameters (capacity, COP and outlet conditions) are finalized.

Air-Side Heat Transfer Coefficient

The airside is assumed to be incompressible and fully developed hydraulically. The relevant conservation equations are continuity, momentum and energy equations. Most air-side heat transfer correlations for different fin geometries are developed for both a single tube and a bank of tubes in crossflow. Webb (1990) suggested the air-side heat transfer correlation for heat exchangers with flat and wavy plate fins on a staggered array of circular tubes as shown in Equations 3.25 and 3.26.

For flat fin,

$$Nu_a = 0.4 \, Gz^{0.73}(s/D_c)^{-0.23}N_R^{0.23} \quad \text{for } Gz \le 25$$

$$Nu_a = 0.53 \, Gz^{0.62}(s/D_c)^{-0.23}N_R^{0.31} \quad \text{for } Gz > 25 \tag{3.25}$$

For wavy fin,

$$Nu_a = 0.5 \, Gz^{0.86}(P_t/D_c)^{0.11}(s/D_c)^{-0.09}(S_d/P_1)^{0.12}(2S_p/P_1)^{-0.34} \quad \text{for } Gz \le 25$$

$$Nu_a = 0.83 \, Gz^{0.76}(P_t/D_c)^{0.13}(s/D_c)^{-0.16}(S_d/P_1)^{0.25}(2S_p/P_1)^{-0.43} \quad \text{for } Gz > 25 \tag{3.26}$$

where
   Gz = Graetz number ($RePrD_H/L$)
   s = space between fins (fin pitch – fin thickness)
   D_c = fin collar outside diameter
   S_d = peak-to-vallet wave depth
   P_l = longitudinal tube pitch
   S_p = one-half wave length
   N_R = number of tube rows

Refrigerant-Side Heat Transfer Coefficient

A correlation is required for the local refrigerant-side heat transfer coefficients. As for the single-phase heat transfer coefficients, Gnielinski's (1976) correlation shown in Equation 3.27 is used for $2300 \le Re \le 5 \times 10^6$.

$$Nu = \frac{(f/8)(Re-1000)Pr}{1+12.7(f/8)^{0.5}(Pr^{2/3}-1)} \tag{3.27}$$

where
   $f = (0.79 \, \ln(Re) - 1.64)^{-2}$

As for the two-phase heat transfer coefficients, many correlations are available. In this section, one preferred boiling and one condensation correlation are introduced, which provided the smallest error with experimental data. Details of boiling heat transfer correlations are described in Section 8.3; details of condensation heat transfer correlations are described in Section 8.6.

Jung et al. (1989b) developed the following flow boiling correlation for horizontal tubes for the binary mixtures R22/R114 and R12/R152a as shown in Equation 3.28,

$$h = C_{me}Fh_l + S/C_{un}h_{pool} \tag{3.28}$$

where

h_l = heat transfer coefficient for the liquid phase flowing alone

F = forced convection enhancement parameter

where

$F = 2.37 \ (0.29 + 1/X_{tt})^{0.85}$

S = forced convection suppression parameter

where

$S = 4048 \ X_{tt}^{1.22} \ B_O^{1.13}$ for $X_{tt} < 1$

$S = 2 - 0.1 \ X_{tt}^{-0.28} \ B_O^{-0.33}$ for $1 < X_{tt} < 5$

$C_{me} = 1 - 0.35 \ |X_v - X_l|^{1.56}$

where

X_l = liquid phase mass fraction

X_v = vapor phase mass fraction

$C_{un} = |1 + (b_2 + b_3)(1 + b_4)| \times (1 + b_5)$

h_{pool} = pool boiling heat transfer coefficient

where

$h_{pool} = h_i/C_{un}$

$h_i = h_a \ h_b/(h_b \ X_l + h_a \ X_v)$

$h_a \ h_b$ = pool boiling heat transfer coefficient of pure refrigerants a and b given by Stephan and Abdelsalam's (1980) correlation

$b_2 = (1 - X_l) \ln |(1.01 - X_l)/(1.01 - X_v)| + X_l \ln (X_l/X_v) + |X_l - X_v|^{1.5}$

$b_3 = 0$ for $X_l \mu \ 0.01$

$b_3 = (X_l/X_v)^{0.1} - 1$ for $X_l < 0.01$

$b_4 = 152 \ (P/P_{ca})^{3.9}$

$b_5 = 0.92 \ |X_l - X_v|^{0.001} + (P/P_{ca})^{0.66}$

$X_l/X_v = 1$ for $X_l = X_v = 0$

where

P_{ca} = critical pressure of more volatile component

Sweeney (1996) and Sweeney and Chato (1996) correlated Kenney et al.'s (1994) data for R407C in a smooth tube using Dobson's correlations,

$$h = 0.7 \left(\frac{G}{300} \right)^{0.3} \left\{ \frac{k_l}{D} 0.023 \, \mathrm{Re}_l^{0.8} \, \mathrm{Pr}_l^{0.4} \left(1 + 2.22 X_{tt}^{-0.89} \right) \right\} \quad \text{for annual flow regime} \quad (3.29)$$

$$h = \left(\frac{G}{300}\right)^{0.3}\left\{\frac{0.23\,\mathrm{Re}_v^{0.12}}{1+1.11X_{tt}^{0.58}}\left(\frac{Ga\,\mathrm{Pr}_l}{Ja_l}\right)+(1-\theta_l/\pi)Nu_f\right\} \text{ for wavy flow regime} \quad (3.30)$$

where

$$Nu_f = 0.0195\,\mathrm{Re}_l^{0.8}\,\mathrm{Pr}_l^{0.4}\,\phi_l(X_{tt})$$

where

$$\phi_l(X_{tt}) = \sqrt{1.376+\frac{C_1}{X_{tt}^{c_2}}}$$

where

$c_1 = 4.12 + 0.48\,\mathrm{Fr}_l - 1.564\,\mathrm{Fr}_l^2,\ c_2 = 4.12 + 0.48\,\mathrm{Fr}_l - 1.564\,\mathrm{Fr}_l^2,\ \text{for } 0 < \mathrm{Fr}_l \leq 0.7$

$c_1 = 7.242,\ c_2 = 1.655\ \text{for } \mathrm{Fr}_l > 0.7$

where

Fr_l = liquid Froude number

θ_l = angle subtended from the top of the tube to the liquid level

$$1-\frac{\theta_l}{\pi} \cong \frac{\arccos(2\alpha-1)}{\pi} \text{ by Jaster and Kosky (1976)}$$

Tubes and Fins

When the segmentation technique is applied for the heat exchanger models, it is assumed that each segment is at one temperature and that the axial conduction is negligible.

Two heat transfer resistances associated with the tube and fin are the resistance due to the tube itself and the resistance due to the contact between the fin and the tube. The contact resistance is estimated from Sheffield's correlation (Sheffield et al., 1989). The fin efficiency (η_f) and overall surface efficiency (η_o) are calculated using the method described by Incropera and Dewitt (1990) as shown in Equations 3.31 and 3.32,

$$\eta_f = \frac{\tanh(mL)}{mL} \quad (3.31)$$

where

$m = (hP/kA_c)^{0.5}$

h = air-side heat transfer coefficient

P = fin perimeter

k = fin thermal conductivity

A_c = fin cross-sectional area

L = fin height

$$\eta_o = 1 - \frac{A_f}{A_t}(1 - \eta_f) \tag{3.32}$$

where

A_f = fin surface area

A_t = total surface area (fin surface area + fin base area)

Overall Heat Transfer Coefficient

The overall heat transfer coefficient (U) is calculated based on a distributed parameter heat exchanger simulation that calculates a local overall heat transfer coefficient from Equation 3.33 using the air-side and refrigerant-side heat transfer coefficients.

$$U = \frac{1}{\dfrac{A_o}{A_i}\dfrac{1}{h_r} + \dfrac{A_o \ln(r_o/r_i)}{2\pi kL} + \dfrac{1}{\eta_o h_a}} \tag{3.33}$$

where

h_r = refrigerant-side heat transfer coefficient

h_a = air-side heat transfer coefficient

r_o = tube outside radius

r_i = tube inside radius

A_o = tube outside surface area

A_i = tube inside surface area

k = thermal conductivity of tube

L = tube length

Refrigerant-side Pressure Drop

Refrigerant-side pressure drop in single-phase regions (superheated gas and sub-cooled liquid) is calculated by using the Blasius friction factor for turbulent in-tube flow (Blasius, 1913) as shown in Equation 3.34,

$$f = \frac{0.0791}{Re^{0.25}} \quad \text{for } 2.1 \times 10^3 < Re < 10^5 \tag{3.34}$$

As for the two-phase pressure drop, many correlations are available. In this section, the most preferred boiling pressure drop and condensation pressure drop correlations are introduced, having the smallest error with experimental data. Details of boiling pressure drop correlations are described in Section 8.4; details of condensation pressure drop correlations are described in Section 8.7.

Jung and Radermacher (1989) studied the boiling pressure drop of R12, R22, R114, R152a and R12/R152a. The authors suggested a modified version of Martinelli-Nelson correlation as shown in Equation 3.35,

$$\Delta P_{tp} = \frac{2f_{f0}G^2L}{D\rho_l}\left[\frac{1}{\Delta x}\int_{x1}^{x2}\phi_{tp}^2 dx\right] \qquad (3.35)$$

where

$f_{fo} = 0.046 \, \text{Re}^{-0.2}$
G = mass flux
D = tube diameter
ρ_l = liquid density

$$\phi_{tp}^2 = 30.78x^{1.323}(1-x)^{0.477}P_r^{-0.7232}$$

where

P_r = reduced pressure (P/P_c)

Sami et al. (1992) suggested using the Lockhart-Martinelli parameters to calculate the frictional pressure gradient during the condensation process based on their measurement on R22/R152a/R114 and R22/R152a/R124 in an enhanced tube. The suggested correlation is shown in Equations 3.36 through 3.40. They correlated the friction factor using the form of Blasius as shown in Equation 3.41,

$$\Delta P = \Delta P_f + \Delta P_m \qquad (3.36)$$

where

ΔP = total two-phase pressure drop
ΔP_f = frictional pressure drop
ΔP_m = pressure drop due to the momentum change

$$\Delta P_m = \frac{G^2}{\rho_l}\left\{\left[\frac{x^2}{\alpha}\left(\frac{\rho_l}{\rho_v}\right)+\frac{(1-x)^2}{(1-\alpha)}\right]_2 - \left[\frac{x^2}{\alpha}\left(\frac{\rho_l}{\rho_v}\right)+\frac{(1-x)^2}{(1-\alpha)}\right]_1\right\} \qquad (3.37)$$

where

G = mass flux
ρ_l = liquid density
ρ_v = vapor density
x = vapor quality
α = void friction

$$\alpha = \left\{ 1 + \left(\frac{\rho_v}{\rho_l} \right) \left(\frac{1-x}{x} \right) \left(0.4 + 0.6 \left[\frac{x\left(\rho_l/\rho_v\right) + 0.4(1-x)}{x + 0.4(1-x)} \right]^{0.5} \right) \right\}^{-1} \qquad (3.38)$$

$$\Delta P_f / \Delta z = \left(\Delta P / \Delta z \right)_v \Phi_v^2 \qquad (3.39)$$

where

Φ_v = Lockhart-Martinelli parameter

$$\left(\frac{\Delta P}{\Delta z} \right)_v = \frac{2 f_v G^2 x^2}{D \rho_v} \qquad (3.40)$$

where

ΔP_v = vapor flow pressure drop

D = tube diameter

f_v = frictional coefficient considering two-phase flow as vapor flow

$$f = 0.6 \, \text{Re}^{-0.21} \qquad (3.41)$$

To investigate the benefit of temperature glide of R407C, different heat exchanger geometries, crossflow, parallel flow and counter flow were examined. The results are shown in Tables 3.1 and 3.2. R407C showed capacity degradation in the parallel flow heat exchange configuration and capacity improvement in the counter flow heat exchange configuration over R22. Performance penalty of R407C was about 4 to 5% for the parallel flow and performance improvement of R407C was about 3 to 4%

TABLE 3.1
Effect of Heat Exchanger Geometry in Cooling Mode

Heat Exchanger Geometry	Condenser Capacity		Evaporator Capacity	
	R22	R407C	R22	R407C
Crossflow	100	100	100	100
Parallel flow	96	92	98	96
Counter flow	103	103	101	103

Air and refrigerant boundary conditions are based on experimental values. The airside heat transfer coefficient, area and fin efficiency are based on the crossflow values.

Source: Judge and Radermacher, 1997.

TABLE 3.2
Effect of Heat Exchanger Geometry in Heating Mode

Heat Exchanger	Condenser Capacity		Evaporator Capacity	
Geometry	R22	R407C	R22	R407C
Crossflow	100	100	100	100
Parallel flow	96	95	99	98
Counter flow	103	104	100	100

Air and refrigerant boundary conditions are based on experimental values. The airside heat transfer coefficient, area and fin efficiency are based on the crossflow values.

Source: Judge and Radermacher, 1997.

for the counter flow as compared to the crossflow configuration. It should be noted that the penalty of parallel flow is a little bit more severe than the benefit of counter flow since the parallel flow encountered a pinch point.

3.10.1.2 Compressor Model

The compressor is the heart of the vapor compression cycle and there are many levels of complexity among the compressor models in the literature. In general, compressor models are classified into four versions: efficiency model, map-based model, lumped parameter model and distributed parameter model. There are six representative compressor efficiencies: overall isentropic efficiency ($\eta_{overall,isen}$), volumetric efficiency (η_{vol}), motor efficiency (η_{mot}), mechanical efficiency (η_{mec}), isentropic efficiency (η_{isen}) and suction gas superheating efficiency ($\eta_{superheat}$) (Shaffer and Lee, 1976). The definition of these six efficiencies is as follows:

$$\eta_{overall,isen} = \frac{\text{ideal isentropic compression work (from shell inlet to outlet)}}{\text{actual motor power}}$$

$$\eta_{vol} = \frac{\text{actual refrigerant mass flow rate}}{\text{ideal refrigerant mass flow rate}}$$

$$\eta_{mot} = \frac{\text{shaft power}}{\text{actual motor power}}$$

$$\eta_{mec} = \frac{\text{indicated power}}{\text{shaft power}}$$

$$\eta_{isen} = \frac{\text{ideal isentropic compression work}}{\text{indicated power}}$$

$$\eta_{superheat} = \frac{\text{ideal isentropic compression work (from shell inlet to outlet)}}{\text{ideal isentropic compression work (from suction port to outlet)}}$$

(3.42)

where

indicated power = work delivered to the gas

An isentropic analysis model is the simplest model and is independent of the physical characteristics of compressors and various types of refrigerants. Typically, two efficiencies, isentropic and volumetric efficiency, are given as inputs in this model. These two efficiencies can be directly obtained from experimental data or varied as a design parameter.

The second model is a polytropic analysis model. Some investigators prefer to use the polytropic efficiency instead of the isentropic efficiency. The polytropic efficiency is defined by Equation 3.43,

$$\eta_{pol} = \frac{\text{polytropic compression work}}{(\text{refrigerant enthalpy change} \times \text{mass flow rate})} \tag{3.43}$$

where the polytropic compression work (W_p) is defined by Equation 3.44,

$$W_p = \frac{n}{n-1} P_{suc} V_{suc} \left[\left(\frac{P_{dis}}{P_{suc}} \right)^{\frac{n-1}{n}} - 1 \right] \tag{3.44}$$

where the polytropic index (n) is defined by Equation 3.45,

$$\frac{n-1}{n} = \frac{(\gamma-1)/\eta_{pol}}{\gamma} \tag{3.45}$$

This compressor model is attractive because it accounts for several nonidealities in a very simple form and is demonstrated by Welsby and Domanski's general compressor model for an air-conditioning and heat pump simulation (Welsby, 1988; Domanski and McLinden, 1992). In this model, the compressor rotates at a constant RPM and does not account for volumetric efficiency. In other words, the volume flow rate though the compressor is constant; the compressor power and discharge state are determined by assuming polytropic compression with a constant polytropic index. The goal of this polytropic analysis can be broken down into two objectives: one is to determine the refrigerant flow rate and the other is to determine the outlet state. The refrigerant flow rate is obtained by multiplying the volumetric efficiency with the theoretical maximum mass flow rate. The volumetric efficiency accounts for re-expansion of the refrigerant in the clearance volume and the density change of the refrigerant prior to entering the compressor.

A map-based model is one of the most frequently used approaches in industry. In this model, the compressor performance is modeled by curve fitting steady state experimental data. One curve fit method is to define the mass flow rate, discharge state and compressor power as functions of the suction conditions and pressure ratio

(Murphy and Goldschmidt, 1985). Although the map-based model is very accurate for a given fluid at steady state, it is not generally applicable to refrigerants and compressors other than the ones upon which it was based. Compressor manufacturers provide coefficients based on ARI Standards for the positive displacement compressor, ARI Standards 520 (1997) and 540 (1999). Revised ARI Standard 540 (1999) used suction- and discharge-dew point temperatures as its test conditions while AREP (1993) and Rice (1993) used mean temperature of bubble and dew temperature for the evaporating and condensing temperatures. Pure refrigerants, azeotropic mixtures and near-azeotropic mixtures are not affected by dew point- and mean temperature-approaches. However, zeotropic mixtures, especially having a large temperature glide, can show differences when these two approaches are compared. Since mean evaporating and condensing temperatures of zeotropic mixtures are higher and lower than dew temperatures at the suction and discharge, respectively, the dew point approach will result in higher COP due to the reduced pressure ratio than the mean temperature approach.

The most complex compressor model is a distributed parameter compressor model. In this model the complete set of multidimensional energy, momentum and continuity equations is somewhat simplified and solved. MacArthur developed the distributed parameter model that deals with seven different state points (MacArthur, 1984). The compression process is assumed to be polytropic. Heat transfer between different compressor components is accounted for by utilizing constant heat transfer coefficients, as is the effect of thermal storage. Judge (1996) developed a distributed parameter compressor model that accounts for valve dynamics and heat exchange between various components of the reciprocating compressor. Other studies using this type of model include: Ng et al. (1976), Recktenwald et al. (1986), Perez-Segarra et al. (1994) and Lio et al. (1994).

From this literature review, it is clear that there is a wide range of methods available for modeling compressors. The simple method, the efficiency model, is most convenient for design point analysis but is of less accuracy over a range of operating conditions. However, this method is a good approach for parametric studies. The distributed parameter model provides better accuracy but needs larger computing time.

3.10.1.3 Expansion Device Model

The purpose of the expansion device is to reduce the pressure of the refrigerant. The temperature of the refrigerant during expansion is also reduced so that the refrigerant can absorb heat in the evaporator. The simplest expansion device model is a thermodynamic expansion model that assumes an isenthalpic expansion with negligible heat loss. The expansion device inlet enthalpy then equals the outlet enthalpy and this model can avoid any hardware dependence.

The four common expansion devices are capillary tubes, short tube orifices (STs), thermostatic expansion valves (TEVs) and metering valves. In actual systems, the capacity and operating mode determine the preferred expansion device. Capillary tubes are used for small capacity cooling applications; STs are used for medium capacity cooling applications; TEVs are used for larger capacity cooling and heating applications. Among the expansion devices, the ST garners attention recently because of its simplicity, low cost and serviceability.

Short Tube Models

Though the geometry of the STs is very simple, the expansion process within the STs is so complex that it is difficult to accurately correlate experimental data from various refrigerants and boundary conditions. Three modeling efforts are introduced as follows:

Mei (1988) developed a ST model based on his experimental data with R22. He concluded that the degree of upstream subcooling and the pressure difference were the major parameters of STs. He then suggested using a general orifice type model but the orifice constant should be a function of those two parameters. Mei reported that his correlation matched experimental data 3 and 4% in average for STs opening and length, respectively.

Aaron and Domanski (1990) developed a ST model based on their experimental data with R22. In their research, they found that the mass flow rate varied only 3 to 8% depending on downstream pressure in "choked" flow condition. The second finding was that the mass flow rate depended upon the degree of subcooling: the mass flow rate increased approximately 23% when the subcooling increased by 8.3°C. They also reported that doubling the length of the ST reduced the mass flow rate by 5% and inlet chamfering increased the mass flow rate by 5 to 25%, while the exit chamfering had no impact on the flow rate. Based on the orifice equation (Equation 3.46), they modeled the ST:

$$\dot{m} = C_c A_c \sqrt{2\rho(P_1 - P_2)} \qquad (3.46)$$

where

C_c = ST chamfer coefficient
A_c = ST cross-sectional flow area
ρ = density of upstream fluid entering ST
P_1 = upstream pressure
P_2 = downstream pressure

Kuehl and Goldschmidt (1992) and Kim and O'Neal (1993 and 1994a) modified the P_2 of the form of the Aaron's equation to correlate their data. Kornhauser (1993) developed a more physically based model with Aaron's experimental data by considering the metastable conditions occurring within the ST. However, this model is applicable only to "choked" flow regions.

Since the correlations developed by Aaron, Kuehl and Kim were based on R22, their correlations must be compared to experimental data with mixtures. Judge (1996) compared his experimental data for R22 and R407C with Aaron and Kim's correlations. He reported that Aaron's model predicted the low side pressure of R22 and R407C within 6 and 11% errors, respectively. He also reported that Kim's model had errors about twice as large as those of Aaron's when it was compared with R22 experimental data.

Most recently, Payne and O'Neal (1998 and 1999) investigated mass flow characteristics of R407C and R410A with and without POE oil through ST orifices 12.7, 19.1 and 25.4 mm in length with diameters of 1.09, 1.34, 1.71 and 1.94 mm. Payne and O'Neal then fitted their data to a previously developed semi-empirical model by Kim and O'Neal (1994a and 1994b), essentially the same form as Equation 3.46. Payne and O'Neal's correlation for R407C and R410A can handle two-phase inlet conditions as shown in Equation 3.47. The effects of oil on the refrigerant mass flow rate was reflected in the two-phase quality correction factor (C_{tp}) and adjusted down stream pressure (P_{2a}),

$$\dot{m} = C_{tp}A_{cs}\sqrt{2\gamma_c\rho(P_1 - P_{2a})}\tag{3.47}$$

where

$\dot{m}$ = mass flow rate [kg/s]
C_{tp} = two-phase quality correction factor
(C_{tp} = 1 for single phase; use Equation 3.48 for two-phase)
γ_c = constant (Table 3.3)
ρ = density of upstream fluid entering ST [kg/m³]
P_1 = upstream pressure [kPa]
P_{2a} = adjusted downstream pressure [kPa] (Equation 3.49)

$$C_{tp} = \left\{\left(1 + a_1x_1\right)\left(1 + a_2LD^{a_3}Y^{a_4Ln(LD)}\right)\right\}^{-1}\tag{3.48}$$

where

$LD = L/D$
L = length of ST [mm]
D = ST diameter [mm]
a_1, a_2, a_3, a_4 = constant (Table 3.4)

$$Y = \frac{x_1}{1 - x_1}\left(\frac{\rho_l}{\rho_v}\right)^{0.5}$$

where

x_1 = upstream vapor quality
ρ_l = saturated liquid density at upstream pressure
ρ_v = saturated vapor density at upstream pressure

$$P_{2a} = P_{sat}\left\{b_1 + b_2PRA^{b_3}LD^{b_4}SUBC^{b_5} + b_6PRA^{b_7} + \right.$$
$$\left. b_8\exp(b_9DR \times LD^{b_{10}}) + b_{11}EVAP\right\}\tag{3.49}$$

TABLE 3.3
Critical Parameters for Equation (3.47)

Critical Parameter	R407C	R410A
P_c	4619.14 kPa	4949.65 kPa
T_c	359.89 K	345.65 K
γ_c	1.2960×10^{10} kg/kPa-h²-m	1.2960×10^{10} kg/kPa-h²-m

Source: Payne & O'Neal, 1998; Payne and O'Neal, 1999.

TABLE 3.4
Coefficients of Correction Factors for Equations (3.48) and (3.49)

	R407C		R410A	
Coefficients	Pure	1.1% POE Oil	Pure	2.2% POE Oil
A_1	−4.45974577	−4.34974577	3.693038038	1.427618112
A_2	10.6946713	10.45457121	0.120175996	0.530751112
A_3	−0.55303036	−0.663120121	0.194241638	−0.365456266
A_4	0.39429366	0.323273661	0.022577667	0.018669938
B_1	0.963034325	0.980538238	0.874805831	1.050104183
B_2	4.286408416	4.957604391	3.131470913	6.305986547
B_3	−0.278235435	−0.309919995	−0.214726407	0.099138818
B_4	−0.043090943	−0.116219951	0.083394737	−0.045626106
B_5	0.916226528	0.906610038	0.901559277	0.958459297
B_6	0.071794702	0.227476573	−0.020574536	−0.254071783
B_7	0.499098698	0.186773583	0.944446846	0.137198955
B_8	−0.208417565	−0.398196082	−0.418400083	−0.276516186
B_9	−0.034680678	−0.030711793	−0.025322802	−0.014589768
B_{10}	1.844061084	1.587754176	2.33507746	2.5121908
B_{11}	−0.09123591	−0.134132834	0.068890172	0.13087558

Source: Payne & O'Neal, 1998; Payne and O'Neal, 1999.

where
P_{sat} = saturated liquid pressure corresponding to upstream temperature [kPa]
$PRA = P_1/P_c$
$DR = D/1.524$
$SUBC = (T_{sat} - T_1)/T_c$
$EVAP = (P_c - P_2)/P_c$
b_1 through b_{10} = constants (Table 3.4)

3.10.1.4 Refrigerant Mixture Properties

The requirements for representation of mixture properties can be satisfied by the REFPROP Version 5 refrigerant properties database that calculates the thermophysical

TABLE 3.5
Equations of State Available in REFPROP

Equation of State	Applicable Fluids	Source
Carling-Starling-DeSantis (CSD)	All of the pure and mixture fluids except ammonia	DeSantis (1976)
Modified Bennedict-Webb-Rubin (MBWR)	13 fluids (R32, R123, R124, R125, R134a, R143A, R152A, R290, R600, R600A, N-C5, I-C5, R236fa)	Bennedict, Webb and Rubin, (1940)
Extended corresponding states (ECS)	Thermodynamic properties of pure and mixture fluids without an MBWR EOS (Equation of State) and transport properties of all fluids except R134a	Huber and Ely (1992) Huber et al. (1992)
Haar-Gallagher (HG)	Ammonia	Haar and Gallagher (1978)

Source: Huber et al., 1996.

properties of 43 pure fluids and their mixtures of up to five components (Huber et al., 1996). REFPROP uses four separate models to calculate thermodynamic properties of pure fluids as shown in Table 3.5.

Among the above equations of state, the ECS model is adapted for the calculation of R22 properties because only CSD and ECS models are available for R22 and the ECS model represents the vapor pressure and saturated liquid density to nearly the uncertainty in experimental data within 0.6%, while the CSD model exhibits systematic deviations that increase as the critical point is approached (McLinden et al., 1993). Uncertainties of transport properties are 10% for thermal conductivity and 15% for viscosity (Huber et al., 1996).

As for the reference state for enthalpy and entropy, the REFPROP database offers three options. Among these, the ASHRAE option is chosen as a convention in this field because it has zero enthalpy and entropy for saturated liquid at –40°C. REFPROP Version 6 is available now with increased accuracy formulations for both pure and mixed refrigerants.

3.10.2 CYCLE MODELS

3.10.2.1 NIST Model: Cycle 7 and Cycle 11

As a serious effort to evaluate the performance of possible alternatives for the existing refrigerants, researchers at NIST developed several cycle simulation models. Initially, McLinden (1987) developed a cycle model Cycle 7 which consists of seven cycle state points as shown in Figure 3.10.2(a).

Cycle 7 then evolved by adding a suction-line heat exchanger and by improving the compressor model, resulting in Cycle 11 which consists of eleven state points as shown in Figure 3.10.2(b) (Domanski and McLinden, 1992).

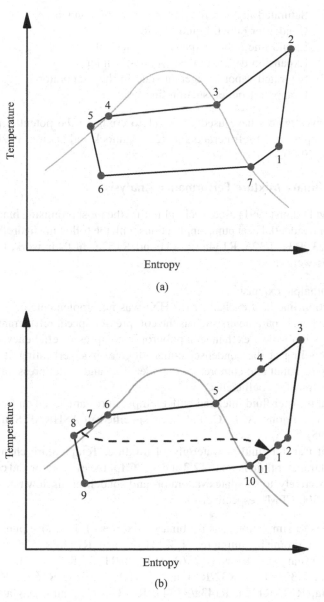

FIGURE 3.10.2 State points of two cycle models. (a) State points of Cycle 7. (b) State points of Cycle 11.

Eleven state points shown in Figure 3.10.2(b) are:

point 1   Suction line outlet, compressor shell inlet
point 2   Refrigerant state in the cylinder before the compression process
point 3   Refrigerant state in the cylinder after the compression process
point 4   Compressor shell outlet, condenser inlet
point 5   Saturated vapor refrigerant state in the condenser

point 6   Saturated liquid refrigerant state in the condenser
point 7   Condenser outlet, liquid line inlet
point 8   Liquid line outlet, expansion device inlet
point 9   Expansion device outlet, evaporator inlet
point 10   Saturated vapor refrigerant state in the evaporator
point 11   Evaporator outlet, suction line inlet

Other investigators have used Cycle 11 to compare the potential performance of possible replacement refrigerants for R22 (Pannock and Didion, 1991; Domanski and Didion, 1993).

3.10.2.2   Binary Mixture Performance Analysis

Pannock and Didion (1991) used Cycle 11 to find the most promising binary zeotropic mixtures for residential heat pump applications with the following hydrofluorocarbons (HFCs): R23, R32, R125, R134a, R143a and R152a. In their work, the following assumptions were used:

- Isenthalpic expansion.
- Suction-line heat exchanger (SLHX) was not implemented.
- Fixed cycle parameters such as the compressor speed, refrigerant pressure drops in the heat exchangers, polytropic compressor efficiency, degree of subcooling at the condenser outlet, degree of superheating at the compressor outlet and compressor cylinder inlet and outlet pressure drop and heat transfer coefficients.
- Heat transfer fluid inlet and outlet temperatures are based on four cooling test conditions, A, B, C, and D, as specified in ASHRAE Standard 116 (1995).
- Heat transfer fluid is water/glycol mixture. Temperature change of the heat transfer fluid is set to 12.2 and 8.2°C for the evaporator and condenser, respectively, to simulate evaporator- and condenser-air flow rate 400 CFM and 800 CFM, respectively.

After a series of simulation runs for binary mixtures of the above pure refrigerants, the two most promising mixtures, R32/R152a and R134a/R152a, were selected among 15 binary mixtures (R23/R32, R23/R125, R23/R143a, R23/R134a, R23/R152a, R32/R125, R32/R143a, R32/R134a, R32/R152a, R125/R143a, R125/R134a, R125/R152a, R143a/R134a, R143a/R152a and R134a/R152a). The R32/R152a mixture showed the best performance. The second and third best performing mixtures were R32/R134a and R125/R134a, respectively.

Figures 3.10.3 and 3.10.4 show the relative volumetric capacity and COP versus the R32 mass fraction to those of R22. Due to the large number of graphs generated by this study, the authors refrained from appending all diagrams and restricted themselves to the best performing mixtures: R32/R152a and R32/R134a. The volumetric capacity of R32/R152a reached that of R22 at 37 and 44% of R32 mass fraction for cooling and heating, respectively. The maximum COP of R32/R152a for cooling and heating was reached at about 40 to 50% of R32 mass fraction. The R32

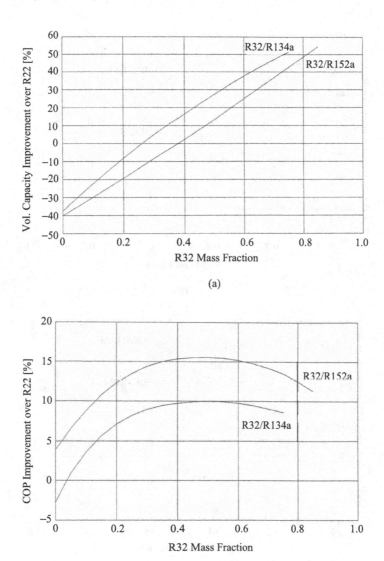

FIGURE 3.10.3 Performance of R32/R152a and R32/R134a at ASHRAE Cooling A conditions (Pannock and Didion, 1991). (a) Relative volumetric capacity. (b) Relative COP.

mass fraction should be about 50% for R32/R152a to have better cooling and heating performance than R22. One interesting result of R32/R152a is that the heating COP has two peak COPs as the R32 mass fraction increases. Pannock and Didion (1991) explained this phenomenon by the temperature glide of the mixture: the first maximum is reached at the point of best glide matching and the COP decreases by the further increase in R32 mass fraction due to the excessive temperature glide of the mixture. However, the temperature glide of the mixture starts to decrease by further increase of the R32 mass fractions, which results in the second peak in the COP by

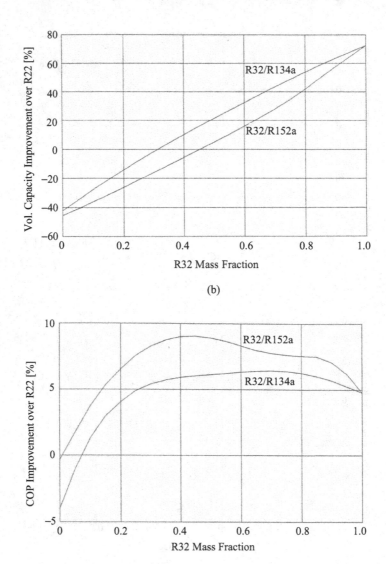

FIGURE 3.10.4 Performance of R32/R152a and R32/R134a at ASHRAE Heating 47S conditions (Pannock and Didion, 1991). (a) Relative volumetric capacity. (b) Relative COP.

better temperature glide matching. However both mixture components of R32/R152a are flammable though it has the lowest GWP out of the 15 considered mixtures.

The volumetric capacity of R32/R152a reached that of R22 at 27 and 30% of R32 mass fraction for cooling and heating, respectively. The maximum COP of R32/R152a for cooling and heating was reached at about 50 to 70% of R32 mass fraction. The R32 mass fraction should be about 50% for R32/R134a to have better cooling and heating performance than R22, the same result with R32/R152a though R32/R134a showed a better volumetric capacity but worse COP than those of

R32/R152a for the same R32 mass fraction. R134a is not flammable, so R32/R134a is definitely less flammable than R32/R152a. Experimental results of these two mixtures are referred to the same study explained in Section 5.1.2.

This investigation of binary zeotropic mixtures showed their potential as alternative refrigerants for R22. It was also found that matching the temperature glide of the refrigerant and air are important to show zeotropic mixtures' potential. In actual residential air conditioners, the indoor heat exchanger can approach cross-counter flow but the outdoor heat exchanger cannot since the number of rows is typically one or two. The temperature glide of mixtures then is not utilized in the outdoor heat exchanger. Another situation to be noted is the refrigerant flow reversal case by changing operating mode from one to another. The temperature glide of mixtures is then not utilized in the cross-counter flow type indoor coil during the heating mode, as will be discussed in Section 5.2.

3.10.2.3 Refrigerant Performance Comparison

Domanski and Didion (1993) also reviewed the performance of nine R22 alternatives, as listed in Table 3.6. The number in the first column of this table designates each refrigerant in Figures 3.10.5 through 3.10.7. Using these nine refrigerants, three cycle options were investigated: drop-in, system modification, and system with SLHX.

In this model analysis, the following assumptions were made:

- Overall compressor polytropic efficiency: 0.7
- Pure crossflow heat exchangers for both evaporator and condenser
- Zero degree of superheating at the evaporator outlet
- Zero degree of subcooling at the condenser outlet
- Fixed pressure drop across heat exchangers (both evaporator and condenser): 34.5 kPa
- Fixed heat transfer fluid inlet and outlet temperatures (Table 3.7)

TABLE 3.6
Refrigerants Investigated

No.	Refrigerant	Mass Fraction [%]	Molecular Mass [kg/kmol]	T_{dew} [°C] at 1 atm	T_{glide} [°C]
0	R22	100	86.5	−40.9	0
1	R32/125	60/40	67.3	−53.1	0
2	R32/125/134a/290	20/55/20/5	86.8	−45.2	8.5
3	R32/125/134a	10/70/20	102.9	−42.4	5.4
4	R290	100	44.1	−40.0	0
5	R32/125/134a	30/10/60	80.1	−36.0	7.6
6	R32/227ea	35/65	94.8	−35.3	20.6
7	R32/134a	30/70	79.2	−34.4	7.3
8	R32/134a	25/75	82.3	−33.1	7.0
9	R134a	100	102.0	−26.2	0

Source: Domanski and Didion, 1993.

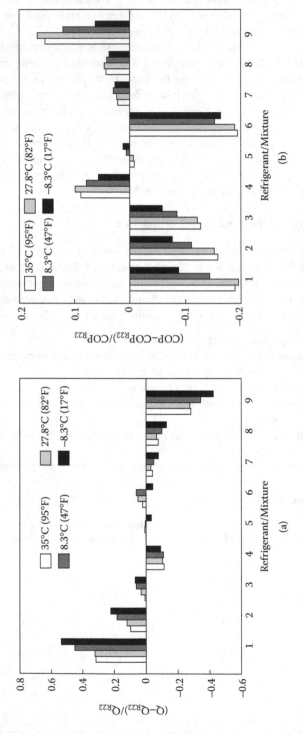

FIGURE 3.10.5 Drop-in performance (Domanski and Didion, 1993). (a) Relative volumetric capacity. (b) Relative COP.

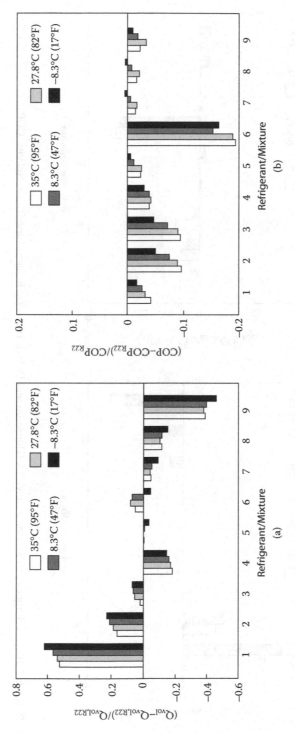

FIGURE 3.10.6 Performance of modified system (Domanski and Didion, 1993). (a) Relative volumetric capacity. (b) Relative COP.

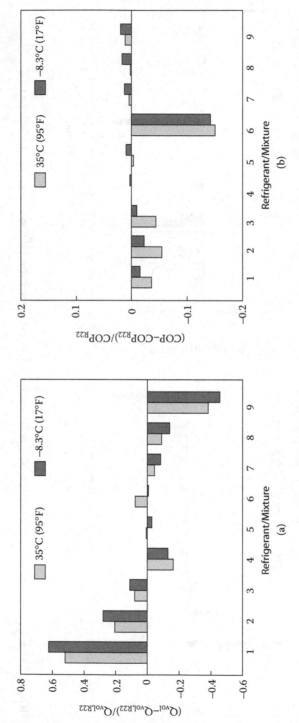

FIGURE 3.10.7 Performance of system with SLHX (Domanski and Didion, 1993). (a) Relative capacity. (b) Relative COP.

TABLE 3.7
Heat Transfer Fluid (Air) Temperatures

Mode	Cooling A	Cooling B	Heating 47S	Heating 17L
Condenser Inlet	35.0°C	27.8°C	21.1°C	21.1°C
Condenser Outlet	43.2°C	37.4°C	32.5°C	28.1°C
Evaporator Inlet	46.7°C	26.7°C	8.3°C	-8.3°C
Evaporator Outlet	14.4°C	13.8°C	2.7°C	-11.3°C

Source: Domanski and Didion, 1993.

In the drop-in performance analysis, it was assumed that there was no system modification except the expansion device. The compressor RPM, displacement volume and heat exchanger areas were fixed as those of the R22 system. Figure 3.10.5 shows drop-in performance results. The refrigerant that has the higher saturation pressure has the higher volumetric capacity. R32/R125 has the highest volumetric capacity and R134a has the lowest volumetric capacity. However, the COP results were reversed: the refrigerant that has higher saturation pressure has the lower COP.

The second analysis was the performance comparison under equal heat exchanger loading as suggested by McLinden and Radermacher (1987). In this analysis, the overall heat capacities of evaporator and condenser (UA_e and UA_c) and average evaporation temperature ($T_{e,avg}$) were selected such that the R22 saturation temperatures at the evaporator and condenser outlet were 7.8 and 46.1°C, respectively, at cooling A test condition. The system was then modified for each refrigerant such that the two constraints of Equations 3.50 and 3.51 were satisfied. Equation 3.50 requires the same heat exchanger area loading and Equation 3.51 requires a specific size of the evaporator.

$$\frac{Q_e}{UA_e + UA_c} = \text{const as R22 value} \tag{3.50}$$

$$T_{e,avg} = \text{const as R22 value} \tag{3.51}$$

Figure 3.10.6 shows the performance of the refrigerants in the modified system. This figure shows that the modified system has a larger change in volumetric capacity than the drop-in case since a larger heat exchanger area is required as the volumetric capacity increases (Equation 3.50). This effect on the volumetric capacity minimized the penalty in COP, so the modified system has less change in COP than the drop-in case. One important result is that the COPs of all nine refrigerants are lower than that of R22.

The third analysis was conducted for the modified system with a SLHX. This analysis was done on cooling for ASHRAE A and the 17L heating conditions. It was further assumed that the SLHX effectiveness was 45%. Figure 3.10.7 shows results of an analysis for the system with SLHX. The COP of the modified system with the

SLHX improved over that without the SLHX. Most refrigerants except R32/R227ea showed a comparable COP with R22 within 5%.

3.10.2.4   ORNL Model: BICYCLE and Mark V

Simplified Cycle Model: BICYCLE

Rice et al. (1993) developed a simplified model, BICYCLE, for modeling design and evaluation of off-design performance of R22 alternative refrigerant mixtures in a vapor compression cycle. In BICYCLE, the following assumptions were used.

- Effects of dehumidification on the heat exchanger UA values are included based upon the sensible heat ratio results of R22.
- UA variation due to the changes in heat exchangers function from cooling to heating mode: −25 to +20%.
- R22 baseline conditions at the compressor suction and discharge: saturated temperatures of 8.9°C/47.8°C.
- Zero degree of superheating at the evaporator outlet.
- Zero degree of subcooling at the condenser outlet.
- A constant total UA value (HX loading) across both heat exchangers (similar to Equation 3.50).
- Counterflow heat exchangers.
- Fixed pressure drops across the heat exchangers (both evaporator and condenser): 34.5 kPa.
- Compressor volumetric efficiency was given as a function of pressure ratio based on the compressor calorimeter tests.
- Compressor isentropic efficiency was given as a function of ambient temperature but was the same for different refrigerants.
- Compressor shell heat loss was 10% of the compressor power input.

Based on the assumptions made above, BICYCLE simulates cycle performance at specified airflow rates (or temperature change across the heat exchanger), heat exchanger UAs and capacity. BICYCLE was verified by comparing its results with results of the detailed model Mark IV and showed an agreement within 4% over a range of cooling and heating conditions. Then the BICYCLE was applied to four different refrigerant mixtures selected from the Alternative Refrigerants Evaluation Program (AREP, 1993) to cover azeotropic mixtures and zeotropic mixtures having small and large temperature glides. Those were one azeotropic mixture, R32/R125 (60/40 wt.%), two small temperature glide mixtures, R32/R134a (30/70 wt.%) and R32/R125/R134a (30/10/60 wt.%), and one large temperature glide mixture, R32/R227ea (35/65 wt.%). For these mixtures, BICYCLE was applied at the same airflow rate, UA values and sensible heat ratios as for R22. Only the compressor displacement was adjusted to obtain the same design capacity.

Results of the calculations are shown in Figure 3.10.8. As intended, all mixtures have the same capacity as R22 at cooling test A conditions. In Figure 3.10.8(b), the COP prediction by Rice et al. (1993) with counterflow heat exchangers are compared to those by Domanski and Didion (1993) as shown by the horizontal lines for

(a)

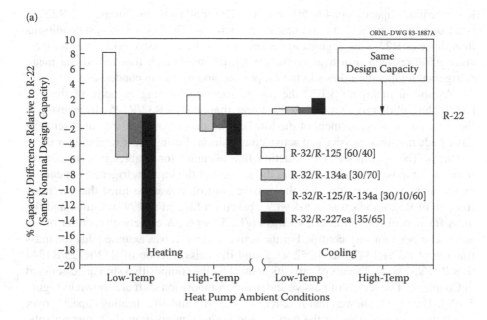

(b)

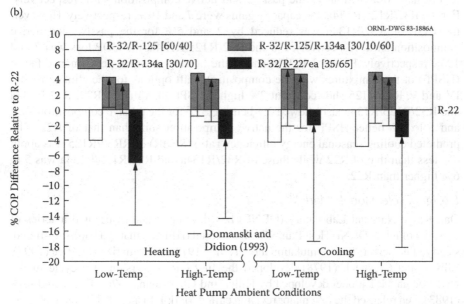

FIGURE 3.10.8 Performance of four refrigerant mixtures (Rice et al., 1993). (a) Relative capacity and COP. (b) Relative heating seasonal performance factor (HSPF).

crossflow heat exchangers with more restrictive conditions (Equation 3.51). As shown in Figure 3.10.8, R32/R125 showed a higher heating capacity especially at the 17L conditions but 4 to 5% lower COP for all test conditions. Two small temperature glide mixtures, R32/R134a and R32/R125/R134a, showed a 2 to 5%

lower heating capacity but 4 to 5% higher COP for all tests conditions. R32/R227ea is the only one that showed lower heating capacity and COPs for more test conditions than those of R22. This degradation is because of the excessive refrigerant temperature glide resulted in a heat exchanger temperature pinch that forced the mean refrigerant temperature down in the evaporator and up in the condenser.

As shown in Figure 3.10.8 the low temperature heating capacity of the low temperature glide mixtures is 4 to 6% lower than that of R32/R125. To investigate the potential for improvement of the low heating performance of the small temperature glide mixtures, passive and active composition shifting was examined for these mixtures. The utilization of the suction line accumulator enables passive control over the composition shift by varying the amount of the liquid refrigerant depending on the ambient temperature. For the passive control, it was assumed that the mass fraction of R32/R134a was 36/64 wt.%, based on Shiflett's (1993) results, and the mass fraction of R32/R125/R134a was 36/12/52 wt.%. A cycle with rectifier offers active composition shift control. For the active control, it was assumed that the mass fraction of R32/R134a was 48/52 wt.% and the mass fraction of R32/R125/R134a was 48/16/36 wt.%. Details of capacity control by the composition shift are discussed in Chapter 7. The results of passive and active composition shift are shown in Figure 3.10.9. The results showed that R32/R134a gained 5 and 16% heating capacity over its original composition for the passive and active composition shift, respectively. For the R32/R125/R134a, the capacity gains were 7 and 19%, respectively. However, the COP of R32/R134a was reduced by 2 and 5% for the passive and active composition shift, respectively. For the R32/R125/R134a, the COP loss was 7 and 19%, respectively. Figure 3.10.9(b) shows the heating seasonal performance factor (HSPF) of three mixtures with the composition shift options for the climate region IV and V. R32/R125 showed about 2% higher HSPF than that of R22. R32/R134a and R32/R125/R134a also showed 2% better HSPF for the passive composition shift and 5 to 8% better HSPF for the active composition shift than that of R22. The predicted cooling seasonal energy efficiency ratio (SEER) of RR32/R125 was about 4% less than that of R22 while those of R32/R134a and R32/R125/R134a was 5 to 6% higher than R22.

Design Cycle Model: Mark V

Oak Ridge National Laboratory (ORNL) has developed and distributed a series of versions of DOE/ORNL Heat Pump Design Model (HPDM) that is a hardware-based, steady-state performance simulation model since 1978 as summarized by Rice (1997). Ellison and Creswick (1978) developed the first ORNL version of a cycle model based on an MIT model developed by Hiller and Glicksman (1976). Fisher and Rice (1983) then released the first major revision called Mark I. Fisher et al. (1988) released Mark III as a PC DOS version. In 1991, Spatz reported using a modified version of the Mark III model to first identify the system potential of R410A, considering transport and thermodynamic property effects on heat exchanger performance. Rice (1987 and 1988) added charge inventory capability and Rice (1991) completed a variable-speed model of the Mark III by adding electronically commutated motors and design parametric capability, called Mark IV. Recently, ORNL released the Mark V version (1994). The latest version of this model includes the variable-speed

(a)

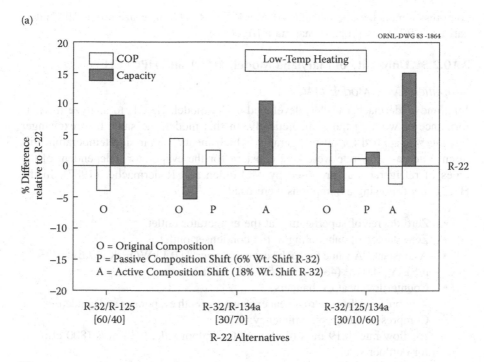

(b)

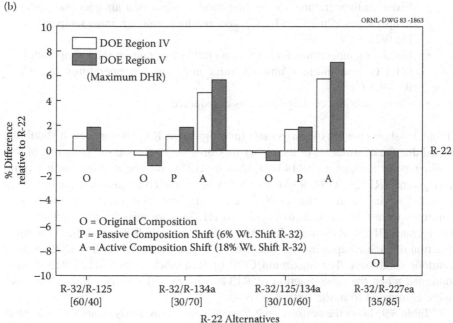

FIGURE 3.10.9 Performance of four refrigerant mixtures with composition shift (Rice et al., 1993a).

compressor model, properties of R410A, R407C, R404A (near-azeotropic HFC alternatives to R22) and propane (natural refrigerant).

3.10.2.5 University of Maryland Model: HAC1 and HPCYCLE

Simplified Cycle Model: HAC1

Jung and Radermacher (1989) developed a UA-model, HAC1, to analyze the performance of working fluids theoretically. In this model, the same heat exchanger loading was applied for all refrigerants, which meant the air-side temperature difference and air flow rate were kept constant for the evaporator independent of the types of refrigerants as proposed by McLinden and Radermacher (1987). In the HAC1, the following assumptions were used.

- Zero degree of superheating at the evaporator outlet.
- Zero degree of subcooling at the condenser outlet.
- A constant UA value for each heat exchanger: U 0.1 kW/m²-K (17.6 Btu/ft²-h°F), A_{ev} 4.0 m² (43 ft²), A_{co} 5.0 m² (54 ft²).
- Counterflow heat exchangers.
- Zero pressure drops across heat exchangers (both evaporator and condenser).
- Compressor isentropic efficiency: 0.7.
- Air flow rate: 0.19 m³/s (400 cfm) for indoor side, 0.38 m³/s (800 cfm) for outdoor side.
- Fixed air temperatures for cooling mode: indoor unit air inlet and outlet temperatures (26.7°C, 11.1°C) and outdoor unit air inlet temperature (35.0/27.8°C).
- Fixed air temperatures for heating mode: indoor unit air inlet temperature (21.1°C) and outdoor unit air inlet and outlet temperatures (8.3°C, 0.6/–8.3°C).
- Only a sensible cooling load was considered.

Table 3.8 shows the list of refrigerants investigated as R22 alternatives. It should be noted that the constituents of the binary mixtures were selected such that one of low boiling point refrigerants (R143a, R32 and R125) and one of high boiling point refrigerants (R152a, R134, R134a and R124) were mixed to minimize the operating pressure deviation from that of R22. The refrigerant components of the ternary mixtures were selected such that R32 and R152a were mixed with nonflammable refrigerants (R134, R134a and R124) to reduce the overall flammability. The mass fraction of each component was determined to maximize the cooling COP at 35°C outdoor conditions. The maximum COP of R143a/R134a and R125/R134a were obtained when the mass fraction of R134a was zero. These two binary mixtures were eliminated from the further analysis.

Table 3.9 shows the performance comparison of pure refrigerants tested at 35°C outdoor conditions. Results indicated that R32 would help to increase the volumetric capacity whereas R124, R134 and R134a would decrease the volumetric capacity. R152a and R134 would help to increase COP, whereas the others would diminish the COP.

TABLE 3.8
List of Refrigerants Investigated

Pure Refrigerant	Binary Refrigerant	Ternary Refrigerant
R32	R32/R134a (30/70 wt.%)	R32/R152a/R134a (40/50/10 wt.%)
R124	R32/R152a (40/60 wt.%)	R32/R152a/R134 (30/40/30 wt.%)
R125	R32/R134 (30/70 wt.%)	R32/R152a/R124 (20/20/60 wt.%)
R134	R32/R124 (30/70 wt.%)	
R134a	R143a/R134a (100/0 wt.%)	
R143a	R143a/R152a (10/90 wt.%)	
R152a	R143a/R124 (30/70 wt.%)	
	R125/R134a (100/0 wt.%)	
	R125/R152a (20/80 wt.%)	
	R125/R124 (50/50 wt.%)	

Source: Radermacher and Jung, 1993.

TABLE 3.9
Performance Comparison of Pure Refrigerants Investigated

Pure Refrigerant	COP	COP Changes [%]	Volumetric Capacity [kJ/m^3]
R22	3.37	0	3735
R32	3.23	−4.2	6227
R124	3.04	−9.8	1382
R125	2.45	−27.4	3305
R134	3.44	2.0	1940
R134a	3.33	−1.2	2342
R143a	2.89	−14.3	3441
R152a	3.56	5.8	2308

Source: Radermacher and Jung, 1993.

The performance of the binary and ternary mixtures is compared in Figure 3.10.10. As shown, four refrigerants (R32, R32/R152a, R32/R134a and R32/R152a/R134a) showed higher volumetric capacity than that of R22 and twelve refrigerants showed higher COP than that of R22. The COP of R32/R152a/R124 was the highest and 14% higher but had 23% lower volumetric capacity than those of R22. R32/R124 was the best binary mixture having 13% higher COP but 10% lower volumetric capacity than those of R22. R32/R152a showed 13% COP and 2% volumetric capacity improvement over R22.

Since R32 mixtures certainly showed better COP than that of R22, only R32 mixtures were further investigated for their seasonal performance factors (SPFs) for both cooling and heating (CSPF and HSPF). It should be noted that only R22 and R32 binary mixtures were examined. CSPF and HSPF were evaluated based on ASHRAE Temperature Bin Method for the unitary air conditioners and heat pumps,

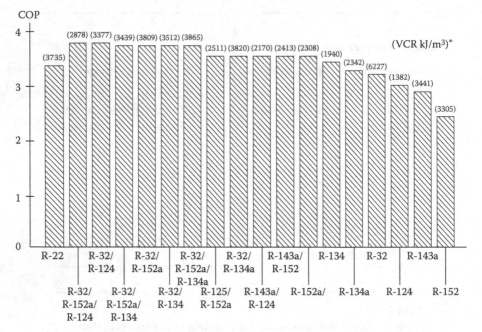

*Numbers in parentheses represent the volumetric capacity (VCR kJ/m³) for each refrigerant

FIGURE 3.10.10 Performance comparison at 35°C outdoor temperature (Radermacher and Jung, 1993).

ASHRAE Standard 116 (1995). The HSPF was evaluated for six climate regions. Results of SPFs are shown in Figure 3.10.11. As shown, R32/R152a exhibited the best SPFs, followed by R32/R134 and R32/R134a.

In order to examine the potential capacity control capability through the concentration shift, it was assumed that the active concentration control could achieve a 20% shift. Then the SPF was evaluated for R32/R152a and R32/R124. As shown in Figure 3.10.12, CSPF was not affected by the concentration shift but the HSPF was increased by 10% for R32/R124 and 6% for R32/R152a.

Detailed Transient Cycle Model: HPCYCLE

Judge (1996) developed a fully implicit, distributed parameter simulation, HPCYCLE, capable of modeling the transient and steady state aspects of an air conditioner and heat pump. For the steady state, the simulation solved the complete continuity, species, energy and momentum equations while transiently only the momentum equation was omitted. The cycle simulation consisted of cycle component simulations. The simulation was applied for both R22 and R32/R134a (30/70 wt.%) to study several different system configurations transiently and at steady state.

From the experimental results, it was found that there was essentially no performance difference between R22 and R32/R134a. In addition, most of the R32/R134a quantities except the refrigerant temperature and the refrigerant concentration were almost identical to R22. Therefore, these two aspects will be explained further.

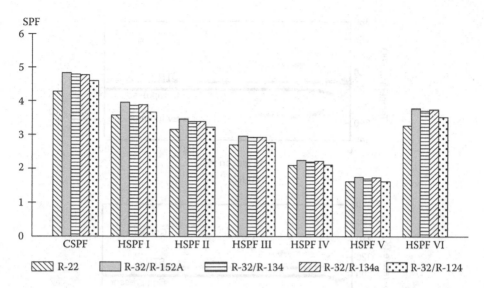

FIGURE 3.10.11 Seasonal performance factor (SPF) comparison (Radermacher and Jung, 1993).

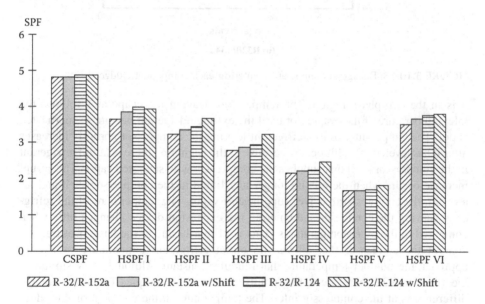

FIGURE 3.10.12 Effects of concentration shift on seasonal performance factor (Radermacher and Jung, 1993).

Figure 3.10.13 shows the refrigerant temperature of R22 and R32/R134a at the inlet of the four-cycle components over time. Two different aspects of the refrigerant temperature of R32/R134a as compared to R22 are as follows: The temperature entering the evaporator closely reproduces the trends of the low side pressure since

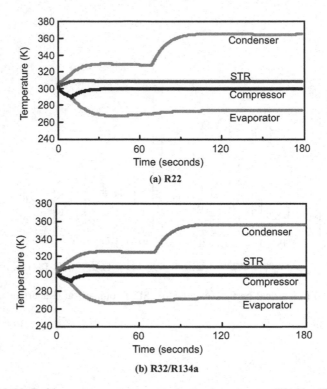

FIGURE 3.10.13 Refrigerant temperature entering each component (Judge, 1996).

it is in the two-phase region. The temperature leaving the evaporator follows the inlet temperature of the evaporator until the exiting refrigerant becomes superheated vapor. The temperature of the refrigerant leaving the compressor abruptly increases after it has leveled off. This occurs because in the beginning there is liquid refrigerant in the compressor shell that supplies the compressor with saturated vapor. The abrupt increase occurs when there is no longer any liquid in the compressor shell and, as a result, the compressor receives superheated vapor. The temperature of the refrigerant leaving the condenser and entering the expansion device changes little when compared to the other temperatures. This is primarily due to the large heat transfer area of the condenser that allows the outlet refrigerant temperature to closely approach the outdoor temperature that remains constant. Although very similar to the plot for R22 there are a few points that make it unique. The most notable difference is at the compressor inlet. The temperature at the compressor inlet does not follow the temperature of the evaporator inlet, as did R22 in the early transients. This is a result of the temperature glide that occurs during the phase change of a zeotropic mixture. Another difference is the considerably lower condenser inlet temperature. This is a consequence of the higher specific heat of R32/R134a. Also noteworthy is the slightly lower evaporator inlet temperature of R32/134a as compared to R22.

The refrigerant concentration is the other quantity that makes R32/R134a unique from R22. Figure 3.10.14 shows the R32 concentration of R32/R134a versus time

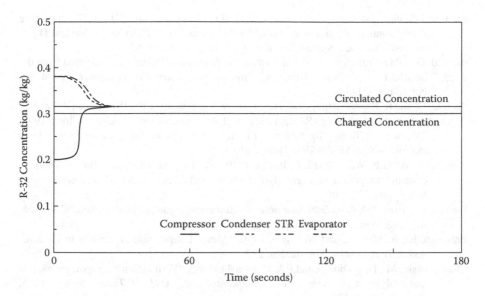

FIGURE 3.10.14 R32 concentration entering each component (Judge, 1996).

at six different locations. Some important aspects of the R32 concentration changes
of R32/R134a are listed below.

Initially, the refrigerant leaving the evaporator and entering the compressor is
depleted of the more volatile component, R32. This is because at start-up, the
refrigerant in the evaporator is in the two-phase region and contains less R32 than
the charged concentration. On the other hand, the vapor, initially in equilibrium with
the refrigerant in the evaporator, is rich in R32 and occupies the condenser. Hence,
the condenser acts like a reservoir of R32 and the evaporator acts like a reservoir
of R134a. As time proceeds, the two-phase, R134a-rich, refrigerant enters the compressor. There the compressor shell separates the liquid from the vapor so that the
compressor compresses only R32-rich vapor. As a consequence the refrigerant leaving the compressor is still relatively rich in R32. However, as the liquid leaves the
evaporator and the amount of liquid in the compressor shell decreases, the R32
concentration leaving the compressor starts to decline. This subsequently causes the
decline in the R32 concentration leaving the condenser. These processes continue
until the circulated concentration is reached at the inlet to each component.

REFERENCES

Aaron, D.A. and P.A. Domanski, 1990, "Experimentation, analysis, and correlation of refrig-
    erant 22 flow through short-tube restrictors," *ASHRAE Transactions*, Vol. 96, Pt. 1,
    pp. 729–742.
Air-Conditioning and Refrigeration Institute, 1997, "Positive Displacement Refrigerant Com-
    pressors," Compressor Units and Condensing Units, ARI Standard 520.
Air-Conditioning and Refrigeration Institute, 1999, "Method for Presentation of Compressor
    Performance Data," ARI Standard 540.

Air-Conditioning and Refrigeration Institute, 1993, "Results from ARI's R22 Alternative Refrigerants Evaluation Program (AREP)," Compressor Calorimeter, System Drop-in, Soft Optimized System Test Reports, Vol. 4, Report #157.

Alefeld, G., 1987, "Efficiency of Compressor Heat Pumps and Refrigerators derived from the Second Law of Thermodynamics," *International Journal of Refrigeration*, Vol. 10, No. 6, pp. 331–341.

Alefeld, G. and R. Radermacher, 1994, *Heat Conversion Systems*, CRC Press, Boca Raton, FL.

American Society of Heating, Refrigerating, and Air-Conditioning Engineers, Inc., 1995, "Methods of Testing for Seasonal Efficiency of Unitary Air-conditioners and Heat Pumps," ASHRAE ANSI/ASHRAE 116-1995.

Bennedict, M., G.B. Webb, and L.C. Rubin, 1940, "An Empirical Equation for the Thermodynamic Properties of Light Hydrocarbons and Their Mixtures," *J. Chem. Phys.*, Vol. 8, pp., 334–348.

Blasius, H., 1913, "Das Ahnlichkeitsgesetz bei Reibungsvorgangen in Flussikeiten," *Forsch. Arb. Ing. Wes.*, 34, Berlin, Germany.

Bowman, R., A. Mueller, and W. Nagle, 1940, "Mean Temperature Difference in Design," *ASME Transactions,* Vol. 62, No. 283.

Chawalowski, M., D.A. Didion, and P.A. Domanski, 1989, "Verification of evaporator models and analysis of performance of an evaporator coil," *ASHRAE Transactions*, Vol. 95, Pt. 1, pp. 1229–1236.

Chen, J.C., 1966, "A Correlation for Boiling Heat Transfer to Saturated Fluids in Vertical Flow," *Ind. Eng. Chem. Proc. Design Dev.*, Vol. 5, No. 3, pp. 322–339.

Chen, S.L., F.M. Gerner, and C.L. Tein, 1987, "General Thin Film Condensation Correlations," *Journal of Experimental Heat Transfer*, Vol. 1, pp. 93–107.

Conde, M.R. and P. Sutter, 1992, "The Simulation of Direct Expansion Evaporator Coils for Air-Source Heat Pumps," International Congress of Refrigeration, Montreal, pp. 1459–1463.

David, S.G., 1994, "Results of Soft Optimized System Tests in ARI's R22 Alternative Refrigerants Evaluation Program," Proceedings of the 1994 International Refrigeration Conference at Purdue, West Lafayette, pp. 7–12.

David, S.G. and M.S. Menzer, 1993, "Results of Compressor Calorimeter Tests in ARI's R22 Alternative Refrigerants Evaluation Program," Proceedings of the 1993 International CFC and Halon Alternatives Conference.

DeSantis, R., F. Gironi, and L. Marrelli, 1976, *Ind. Eng. Chem. Fund.*, Vol. 15: 183.

Dobson, M.K., J.C. Chato, D.K. Hinde, and S.P. Wang, 1994, "Experimental Evaluation of Internal Condensation of Refrigerants R12 and R134a," *ASHRAE Transactions*, Vol. 100, Pt. 1, pp. 744–754.

Dobson, M.K., J.C. Chato, J.C. Wattelet, J.A. Gaibel, M. Ponchner, P.J. Kenney, R.L. Shimon, T.C. Villaneuva, N.L. Rhines, K.A. Sweeney, D.G. Allen, and T.T. Hershberger, 1994, "Heat Transfer and Flow Regimes during Condensation in Horizontal Tubes," ACRC Report 57.

Dobson, M.K. and J.C. Chato, 1998, "Condensation in Smooth Horizontal Tubes," *Journal of Heat Transfer*, Vol. 120, pp. 193–213.

Domanski, P.A., 1989, "HVSIM — An evaporator simulation model accounting for refrigerant and one dimensional air distribution," NISTIR 89-4133.

Domanski, P.A., 1995, "Theoretical Evaluation of the Vapor Compression Cycle With a Liquid-Line/Suction-Line Heat Exchanger, Economizer, and Ejector," NIST Report, NISTIR 5606.

Domanski, P.A. and M.O. McLinden, 1992, "A Simplified Cycle Simulation Model for the Performance Rating of Refrigerants and Refrigerant Mixtures," *International Journal of Refrigeration*, Vol. 15, No. 2, pp. 81–88.

Domanski, P.A. and D.A. Didion, 1993, "Thermodynamic Evaluation of R-22 Alternative Refrigerants and Refrigerant Mixtures," *ASHRAE Transactions*, Vol. 99, Pt. 2, pp. 636–648.

Ellison, R.D. and F.A. Creswick, 1978, "A Computer Simulation of Steady State Performance of Air-to-Air Heat Pumps," ORNL/Con-16.

Fischer, S.K. and C.K. Rice, 1983, "The Oak Ridge Heat Pump Models: I. A Steady-State Computer Design Model for Air-To-Air Heat Pumps," ORNL/CON-80/Rl.

Fisher, S.K. et al., 1988, "The Oak Ridge Heat Pump Model: Mark III Version Program Documentation," ORNL/TM-10192.

Gnielinski, V., 1976, "New Equation for Heat and Mass Transfer in Turbulent Pipe and Channel Flow," *Int. Chemical Engineer*, Vol. 16, pp. 359–368.

Gray, D.L. and R.L. Webb, 1986, "Heat Transfer and Friction Correlations for Plate Finned-Tube Heat Exchangers having Plain Fins," Proc. of Eighth Int. Heat Transfer Conference, San Francisco, pp. 123–131.

Gungor, K.E. and R.H.S. Winterton, 1986, *International Journal of Heat and Mass Transfer*, Vol. 29, No. 3, pp. 351–358.

Haar, L. and J. Gallagher, 1978, "Thermodynamic Properties of Ammonia," *J. Physical Chemical Ref. Data*, Vol. 7, pp. 635–792.

Haar, L., J.S. Gallagher, and G.S. Kell, 1984, *NBS/NRC Steam Tables*, Hemisphere Publishing Co., New York, pp. 301–316 and 635–792.

Herold, K.E., 1989, "Performance Limits for Thermodynamic Cycles," Proceedings of ASME Winter Annual Meeting, AES-Vol. 8, pp. 41–45.

Herold, K.E., R. Radermacher, and S.A. Klein, 1996, *Absorption Chillers and Heat Pumps*, CRC Press, Boca Raton, FL.

Hiller, C.C. and L.R. Glicksman, 1976, "Improving Heat pump Performance via Compressor Capacity Control-Analysis and Test," Vol. 1 & 2, MIT Laboratory Report No. MIT-EL 76-001.

Huber, M.L., D.G. Friend, and J.F. Ely, 1992, "Prediction of the Thermal Conductivity of Refrigerants and Refrigerant Mixtures," *Fluid Phase Equilibria*, Vol. 80, pp. 249–261.

Huber, M.L. and J.F. Ely, 1992, "Prediction of the Viscosity of Refrigerants and Refrigerant Mixtures," *Fluid Phase Equilibria*, Vol. 80, pp. 239–248.

Hwang, Y. and R. Radermacher, 1998, "Emerging Refrigerants," Proceedings of 1998 IIR Conference, pp. 1.1–1.14.

Incropera, F.P. and D.P. Dewitt, 1990, *Fundamentals of Heat and Mass Transfer*, John Wiley & Sons, New York, NY, pp. 645–678.

Jaster, H. and P.G. Kosky, 1976, "Condensation in a Mixed Flow Regime," *International Journal of Heat and Mass Transfer*, Vol. 19, pp. 95–99.

Judge, J., 1996, "A Transient and Steady State Study of Pure and Mixed Refrigerants in a Residential Heat Pump," Ph.D. Dissertation, University of Maryland, College Park, MD.

Judge, J. and R. Radermacher, 1997, "A Heat Exchanger Model for Mixtures and Pure Refrigerant Cycle Simulations," *International Journal of Refrigeration*, Vol. 20, No. 4, pp. 244–255.

Jung, D.S., M. McLinden, R. Radermacher, and D.A. Didion, 1989, "A Study of Flow Boiling Heat Transfer with Refrigerant Mixtures," *International Journal of Heat Transfer*, Vol. 32, No. 9, pp. 1751–1764.

Jung, D.S. and R. Radermacher, 1989, "Prediction of Pressure Drop during Horizontal Annular Flow Boiling of Pure and Mixed Refrigerant," *International Journal of Heat Transfer*, Vol. 32, No. 12, pp. 2435–2446.

Jung, D.S., R. Radermacher, and D.A. Didion, 1989, "Horizontal Flow Boiling Heat Transfer Experiments with a Mixture of R22/R114," *International Journal of Heat Transfer*, Vol. 32, No. 1, pp. 131–145.

Kandlikar, S.G., 1991, "Correlating Flow Boiling Heat Transfer Data in Binary Systems," Proceedings of ASME National Heat Transfer Conference.

Kenney, P. et al., 1994, "Condensation of a Zeotrophic Refrigerant R-32/R-125/R-134a (23%/25%/52%) in a Horizontal Tube." ACRC Report. TR-062.

Kim, Y. and D.L. O'Neal, 1993, "An experimental study of two-phase flow of HFC-134a through short tube orifices," *ASHRAE Transactions*, Vol. 98, Pt. 1, pp. 122–124.

Kim, Y. and D.L. O'Neal, 1994a, "Two-phase flow of Refrigerant-22 through short tube orifices," *ASHRAE Transactions*, Vol. 100, Pt. 1, pp. 323–334.

Kim, Y. and D.L. O'Neal, 1994b, "A semi-empirical model of two-phase flow of HFC-134a through short tube orifices," AES Vol. 29, ASME Winter Annual Meeting.

Kornhauser, A.A., 1993, "Physically Realistic Closed Form Modes for Flashing Flow in Short Tubes," Proceedings of the 28th Inter-Society Energy Conversion Engineering Conference.

Kuehl, S. and V.W. Goldschmidt, 1992, "Flow of R-22 through short tube restrictors," *ASHRAE Transactions*, Vol. 78, pp. 59–64.

Lio, M., A. Doria, and M. Bucciarelli, 1994, "Numerical Analysis of the Dynamics of Reed Valves Taking into Account the Acoustic Coupling with the Fluid," Purdue Compressor Conference, Vol. 2, pp. 229–234.

MacArthur, J.W., 1984, "Theoretical analysis of the dynamic interactions of vapor compression heat pumps," *Energy Conversion Management*, Vol. 24, pp. 49–66.

McLinden, M.O., 1987, "Theoretical Vapor Compression Cycle Model, Cycle 7," National Bureau of Standards, Gaithersburg, MD.

McLinden, M.O. and R. Radermacher, 1987, "Methods for Comparing the Performance of Pure and Mixed Refrigerants in the Vapor Compression Cycle," *International Journal of Refrigeration*, Vol. 10, No. 6, pp. 318–325.

McLinden, M.O., J.G. Gallagher, H.L. Huber, and G. Horrison, 1993, Refprop Refrigerant Properties Database: Capabilities, Limitations, and Future Directions, 1982, Proceedings of ASHRAE/NIST, Refrigerant Conference, Gaithersburg, MD.

Mei, V.C., 1988, "Short-Tube Refrigerant Flow Restrictors," *ASHRAE Transactions*, Vol. 88, Pt. 2, pp. 157–169.

Murphy, W.E. and Goldschmidt, V.W., 1985, "Cyclic Characteristics of a typical Residential Air Conditioner-Modeling of Start-Up Transients," *ASHRAE Transactions*, Vol. 92, Pt. 1A, pp. 186–202.

Ng, E., A. Tramschek, and J. Maclaren, 1976, "Computer Simulation of Reciprocating Compressor Using a Real Gas Equation of State," *International Journal of Refrigeration*, Vol. 15, pp. 33–42.

Nyers, J. and G. Stoyan, 1994, "A Dynamical Model Adequate for Controlling the Evaporator of a Heat Pump," *Intl. J. of Refrigeration,* Vol. 17, p. 2.

ORNL Website, 1998, DOE/ORNL Heat Pump Design Model.

Oskarsson, S.P., K.I. Krakow, and S. Lin, 1990, "Evaporator models for operation with dry, wet, and frosted finned surfaces Part II: Evaporator Models and Verification," *ASHRAE Transactions*, Vol. 96, Pt. 1, pp. 373–380.

Pannock, J. and D. Didion, 1991, "The Performance of Chlorine-Free Binary Zeotropic Refrigerant Mixtures in a Heat Pump," NISTIR 4748, National Institute of Standards and Technology, Gaithersburg, MD.

Payne, W.V. and D.L. O'Neal, 1998, "Mass Flow Characteristics of R407C Through Short Tube Orifices," *ASHRAE Transactions*, Vol. 104, Part 1A, pp. 197–209.

Payne, W.V. and D.L. O'Neal, 1999, "Multiphase Flow of Refrigerant R410A Through Short Tube Orifices," *ASHRAE Transactions*, Vol. 105, Part 2.

Perez-Segarra, C.D., F. Escanes, and A. Oliva, 1994, "Numerical Study of the Thermal and Fluid-Dynamic Behavior of Reciprocating Compressors," Purdue Compressor Conference Proceedings, pp. 145–150.

Rabas, T.J. and P.W. Eckels, 1985, "Heat Transfer and Pressure Drop Performance of Segmented Extended Surface Tube Bundles," ASME Journal of Heat Transfer, Vol. 75, pp. 1–8.

Radermacher, R. and D.S. Jung, 1993, "Theoretical Analysis of Replacement Refrigerants for R-22 for Residential Uses," ASHRAE Transactions, Vol. 99, Pt. 1, pp. 333–343.

Recktenwald, G.W., J.W. Ramsey, and S.V. Patankar, 1986, "Predictions of Heat Transfer in Compressor Cylinders," Purdue Compressor Conference Proceedings, West Lafayette, pp. 159–174.

Rice, C.K., 1987, "The Effect of Void Fraction Correlation and Heat Flux Assumption on Refrigerant Charge Inventory Predictions," ASHRAE Transactions, Vol. 93, Pt. 1, pp. 341–367.

Rice, C.K., 1988, "Capacity Modulation Component Characterization and Design Tool Development," Proceedings of the 2nd DOE/ORNL Heat Pump Conference: Research and Development on Heat Pumps for Space Conditioning Applications, CONF-8804100, Washington D.C., pp.23–33.

Rice, C.K., 1991, The ORNL Modulating Heat Pump Design Tool — User's Guide, ORNL/CON-343.

Rice, C.K., 1993a, "Influence of HX Size and Augmentation on Performance Potential of Mixtures in Air-To-Air Heat Pumps," ASHRAE Transactions, Vol. 99, Pt. 2, pp. 665–679.

Rice, C.K., 1993b, "Compressor Calorimeter Performance of Refrigerant Blends-Comparative Methods and Results for a Refrigerator/Freezer Application," ASHRAE Transactions, Vol. 99, Pt. 1, pp. 1447–1466.

Rice, C.K., L.S. Wright, and P.K. Bansal, 1993, "Thermodynamic Evaluation Model for R22 Alternatives in Heat Pumps-Initial Results and Comparisons," 2nd International Conference on Heat Pumps in Cold Climates, New Brunswick, Canada, Canada Research, pp. 81–96.

Rice, C.K. and W.L. Jackson, 1994, "PUREZ — The Mark V ORNL Heat Pump Design Model for Chlorine-Free, Pure and Near-Azeotropic Refrigerant Alternatives," Preliminary Documentation Package, Oak Ridge National Laboratory.

Rice, C.K., 1997, "DOE/ORNL Heat Pump Design Model, Overview and Application to R22 Alternatives," 3rd International Conference on Heat Pumps in Cold Climates, Nova Scotia, Canada.

Sami, S.M., J. Schnotale, and J.G. Smale, 1992, "Prediction of the Heat Transfer Characteristics of R22/R152a/R114 and R22/R152a/R124," ASHRAE Transactions, Vol. 98, Pt. 2, pp. 51–58.

Sand, J.R., E.A. Vineyard, and R.J. Nowark, 1990, "Experimental Performance of Ozone Safe Alternative Refrigerants," ASHRAE Transactions, Vol. 96, Pt. 2, pp. 173–182.

Shaffer, R.W. and W.D. Lee, 1976, "Energy Consumption in Hermetic Refrigerator Compressors," Proceedings of the 1976 Purdue Compressor Technology Conference at Purdue, West Lafayette.

Shah, M.M., 1979, "A General Correlation for Heat Transfer During Film Condensation Inside Pipes," Journal of Heat and Mass Transfer, Vol. 22, pp. 547–556.

Shah, M.M., 1982, "Chart Correlation for Saturated Boiling Heat Transfer: Equations and Further Studies," ASHRAE Transactions, Vol. 88, Pt. 1, pp. 185–196.

Sheffield, J.W., R.A. Wood, and H.J. Sauer, 1989, "Experimental Investigation of Thermal Conductance of Finned Tube Contacts," Experimental Thermal and Fluid Science, Elsevier Science Publishing Co. Inc., New York, NY, pp. 107–121.

Shiflett, M.B., 1993, "HCFC-22 Alternatives for Air Conditioning and Heat Pumps," ASHRAE Seminar Presentation, DuPont Fluorochemicals, Wilmington, DE.

Spatz, M.W., 1991, "Performance of Alternative Refrigerants From a System Perspective," Proceedings of International CFC and Halon Alternative Conference, Baltimore, MD, pp. 352–363.

Stephan, K. and M. Abdelsalam, 1980, "Heat Transfer Correlations for Natural Convection Boiling," *International Journal of Heat Transfer,* Vol. 23.

Sweeny, K.A., 1996, "The Heat Transfer and Pressure Drop Behavior of a Zeotropic Refrigerant Mixture in a Microfin Tube," M.S. Thesis, University of Illinois at Urbana-Champaign.

Sweeny, K.A. and J.C. Chato, 1996, "The Heat Transfer and Pressure Drop Behavior of a Zeotropic Refrigerant Mixture in a Microfin Tube," ACRC Technical Report 95, University of Illinois at Urbana-Champaign.

Tandon, T.N., H.K. Varma, and C.P. Gupta, 1986, "Generalized Correlation for Condensation of Binary Mixtures Inside a Horizontal Tube," *International Journal of Refrigeration,* Vol. 9, pp. 134–136.

Traviss, D.P., W.M. Rohsenow, and A.B. Baron, 1973, "Forced-Convection Condensation Inside Tubes: A Heat Transfer Equation for Condenser Design," *ASHRAE Transactions,* Vol. 79, Pt. 1, pp. 157–165.

Vineyard, E.A., J.R. Sand, and R.J. Nowark, 1989, "Experimental Performance of Ozone-Safe, Alternative Refrigerants," *ASHRAE Technical Data Bulletin,* Vol. 6, No. 1, pp. 66–75.

Vineyard, E.A., J.R. Sand, and T.G. Statt, 1989, "Selection of Ozone-Safe, Nonazeotropic Refrigerant Mixture for Capacity Modulation in Residential Heat Pumps," *ASHRAE Transactions,* Vol. 95, Pt. 1, pp. 34–46.

Webb, R.L., 1990, "Airside Heat Transfer Correlations for Flat and Wavy Plate Fin-and-Tube Geometries," *ASHRAE Transactions,* Vol. 96, Pt. 2, pp. 445–449.

Welsby, P., S. Devotta, and P.J. Diggory, 1988, "Steady- and Dynamic-State Simulations of Heat Pumps, Part II: Modeling of a Motor Driven Water-to-Water Heat Pump," *Applied Energy,* Vol. 31, pp. 239–262.

4 Methods for Improving the Cycle Efficiency

Several options for the incremental efficiency improvement of the vapor compression cycle are available. In addition, based especially on the understanding of the refrigerant properties and how they affect cycle performance, very significant additional improvements are available when the cycle is modified and new components are added. Three measures are discussed in this chapter, the suction line to liquid line heat exchanger, the economizer and the isentropic expansion device. While these devices also benefit cycles with pure fluids, they have unique effects for mixtures when considered in conjunction with the temperature glide. For mixtures alone, one additional option exists — the three-path evaporator.

4.1 MEASURES OF INCREMENTAL EFFICIENCY IMPROVEMENT

One obvious concern in vapor compression systems is the pressure loss in the connecting piping. Significant efficiency improvements are possible when the pressure drop is minimized. Pipe diameters must be selected so that the pressure drop is small, even though larger pipes contribute to an increased first cost. In addition, especially in the suction pipe, the vapor velocity must be sufficiently high to ensure proper oil return to the compressor. The pressure drop within the heat exchangers should also be minimized. But here again a trade-off exists: Low pressure drop means large tube diameters and low fluid velocities. These in turn lead to poor heat transfer coefficients. Lastly, the flow channels within the compressor and especially the suction and discharge valves (if used in a given compressor type) can contribute significantly to pressure drop and greatly influence overall compressor performance. While a reduction in pressure drop is very desirable, there can be a number of competing effects that limit the level of reduction.

Another measure to improve cycle performance is avoiding evaporator superheat. When the evaporator outlet is saturated vapor, the entire evaporator surface area is used for the purpose of evaporating refrigerant. This occurs at the lowest temperature in the system and at very good heat transfer coefficients. However, some superheat is required to ensure that the compressor is not exposed to any liquid that can be entrained in the suction vapor. The vapor is superheated within the evaporator using valuable surface area. The superheating process absorbs much less heat for a given evaporator area than evaporation because the specific heat capacity is low for superheating, the heat transfer coefficient is low and the average temperature difference is smaller than for evaporation. When superheating is required, the evaporator saturation

135

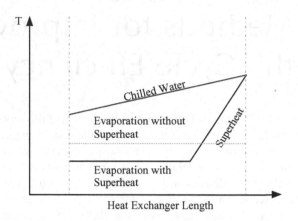

FIGURE 4.1.1 The need for superheat can lower the evaporation temperature.

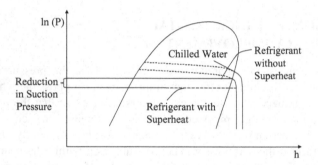

FIGURE 4.1.2 The need for superheat can require a lower suction pressure and thus lower compressor capacity.

temperature is pushed to a somewhat lower level. This is described in Figures 4.1.1 and 4.1.2 for water chilling.

It should be noted that superheating also increases compressor work per unit mass of refrigerant (explained in more detail in the following chapter) and reduces the compressor volumetric capacity. Since the vapor density is also lower, the compressor pumps a smaller refrigerant mass flow rate. This reduces the cooling capacity of the system.

For refrigerant mixtures, a temperature glide exists and the penalty for super-heating remains essentially the same. The influence of the reduced temperature difference can be aggravated by the glide. The argument of the reduced heat transfer coefficient for superheating still applies.

A similar but somewhat more complex situation exists at the condenser outlet. Any subcooling reduces the condenser heat transfer area that is available for con-densation and increases the condenser saturation temperature. Alternatively, sub-cooling benefits the evaporator capacity by reducing the enthalpy at the inlet to the evaporator. The designer must find a trade-off between these two competing effects.

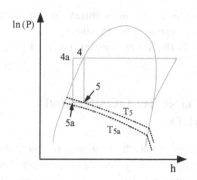

FIGURE 4.1.3 Effects of subcooling.

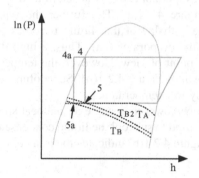

FIGURE 4.1.4 Effects of subcooling for two refrigerant mixtures having different isotherms.

When zeotropic mixtures are used, the subcooling has two effects on the evaporator inlet conditions. As with pure fluids, subcooling reduces the enthalpy and increases the capacity of the evaporating fluid. At the same time, as a result of the glide, the inlet temperature to the evaporator is reduced, as shown in Figure 4.1.3. With increased subcooling from 4 to 4a, the evaporator inlet state point changes from 5 to 5a. As can be seen in Figure 4.1.3, the temperature T_{5a} is lower than T_5. However, depending on the slope of the two-phase isotherm in this portion of the two-phase region, the decreased inlet temperature effect can be more or less pronounced. Figure 4.1.4 shows examples of isotherms for two different refrigerant mixtures, A and B. The isotherm T_A of refrigerant mixture A has an almost horizontal slope in the low quality region close to the saturated liquid line. T_B and T_{B2} of mixture B have a relatively steep slope. Thus, the effect of subcooling on reducing the evaporator inlet temperature as shown in Figure 4.1.3 is much less pronounced for a mixture that exhibits an isotherm of type T_A. There is essentially no difference in temperature between points 5a and 5. For an isotherm of type T_B, however, point 5a is located on the lower temperature isotherm T_B while point 5 lies on T_{B2}, $T_B < T_{B2}$.

This reduction of the evaporator inlet temperature has interesting consequences. The suction pressure can be increased slightly in order to maintain a constant evaporator capacity as long as the glide is still well matched. The compressor power consumption is reduced. This proposal assumes that an increase in subcooling does not push up the condenser pressure to the degree that it overcompensates for the benefit in the evaporator. If the subcooling enhances a glide mismatch, the performance will degrade.

4.2 THE SUCTION LINE TO LIQUID LINE HEAT EXCHANGER

The following consideration focuses first on pure refrigerants. Figure 4.2.1(a) shows a schematic of a vapor compression cycle with a suction to liquid line heat exchanger and Figure 4.2.1(b) the corresponding $\ln(P)$-h diagram. The purpose of the heat exchanger is to increase the evaporator capacity. When saturated liquid leaves the condenser at point 4, the evaporator capacity is determined by the enthalpy difference $h_1 - h_5$ as shown in Figure 4.2.1(a). The further the liquid condensate can be subcooled, the lower the enthalpy of that liquid becomes (4a) and, therefore, the enthalpy at the inlet of the evaporator (5a). Thus, a high degree of subcooling is very desirable from this point of view. However, the temperature T_{4a} is lower than the temperature T_4, as seen in Figure 4.2.1(b). Subcooling requires that a heat sink is available at sufficiently low temperatures.

One heat sink is always lower than any external heat sink — the suction vapor. The suction vapor can be used to subcool the liquid condensate as indicated in Figure 4.2.1(a). The arrow in Figure 4.2.1(b) indicates the transfer of heat, i.e., the enthalpy

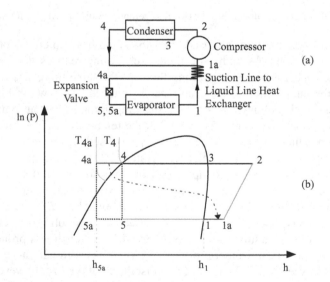

FIGURE 4.2.1 (a) Schematic of vapor compression cycle with suction line to liquid line heat exchanger. (b) The state points according to (a) on an $\ln(P)$-h diagram. The arrow indicates the direction of heat transfer.

difference h_4 to h_{4a} from the subcooled liquid to the suction gas $(h_{1a} - h_1)$. Thus, the suction gas is superheated and its enthalpy increased by

$$q_{slhx} = h_4 - h_{4a} = h_{1a} - h_1 \qquad (4.1)$$

while the inlet enthalpy to the evaporator is lowered by the same q_{slhx}. This transfer of heat enables one to transfer cooling capacity de facto from the evaporator saturation temperature level via the suction heat exchange to the temperature level of the (warmer) suction gas as it is being superheated.

This benefit comes with a challenge. The suction gas is superheated and the state point for the superheated gas moves into a region of the $\ln(P)$-h diagram where the isentropes become flatter and the compressor work input increases. Thus, the benefit of the suction to liquid line heat exchanger depends strongly on whether or not the benefit outweighs the penalty of increased compressor work. This strongly depends on the working fluid used.

As a rule, the following consideration applies: The higher the specific heat capacity of the refrigerant molecules, the higher the benefit of the suction to liquid line heat exchange. A high specific heat capacity of the refrigerant has two effects: The temperature increase during the compression process is low, the compressor compresses a gas that heats up less and the compressor work does not increase as much because of the lack of superheat. However, during the expansion process a relatively large amount of refrigerant must evaporate to cool the remaining liquid to the evaporator saturation temperature.

This situation is better understood when the $\ln(P)$-h diagram for a refrigerant of high specific heat capacity is compared to a refrigerant of low specific heat. Figure 4.2.2 shows two diagrams side by side; (a) shows the diagram of a low capacity refrigerant, (b) that of a high heat capacity fluid. Obviously, the vapor dome is much tilted more to higher enthalpies for the high heat capacity fluid as shown in Figure 4.2.2(b). As a consequence, the expansion process produces much more vapor and

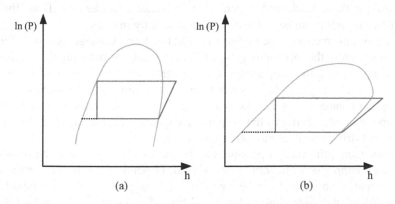

FIGURE 4.2.2 (a) $\ln(P)$-h diagram for a refrigerant of low specific heat capacity. (b) $\ln(P)$-h diagram for a refrigerant of large specific heat capacity.

less liquid. The evaporator capacity is reduced considerably. Furthermore, the compression process stays very close to the saturated vapor line and the compressed refrigerant vapor is almost not superheated at all. In fact, for some refrigerants, the state point of the compressed vapor can be located in the two-phase range. In this case, the penalty in terms of increased compressor work is small or even desirable to avoid a two-phase condition at the compressor outlet. The benefit of improved evaporator capacity is very important because otherwise the refrigerant would not be very efficient.

For refrigerants of low specific heat capacity, the suction to liquid line heat exchanger can have the opposite effect. Examples for fluids of low specific heat capacity are ammonia, R22, R32 and other fluids with a small number of atoms in the molecule. Refrigerants of high specific heat capacity are R123, R134a and other such molecules with a large number of atoms within the molecule. When the number of atoms is small, any addition of energy can only be transferred into translational movement, increasing the temperature. When the number of atoms in a molecule is large, additional energy is first transferred into internal degrees of freedom, such as oscillations and rotations of the atoms within the molecule, and very little energy into the translational movement. The resulting increase in temperature is small.

Figure 4.2.3 shows a plot of COP improvement when a suction to liquid line heat exchanger is used with three levels of heat exchange efficiency, $\eta_{hx} = 0.0$ (baseline), $\eta_{hx} = 0.5$ and $\eta_{hx} = 1.0$. All COP values refer to the Carnot COP and all values are evaluated at the same reduced temperature. The refrigerants on the x-axis are sorted according to their vapor specific heat capacity beginning with the lowest specific heat capacity. For low heat capacity fluids, the suction to liquid line heat exchanger has an obvious detrimental effect. This is followed by a range where there is no particular benefit and the higher specific heat capacity of the refrigerant becomes more pronounced in the improvement resulting from the heat exchanger. It should be noted that the fluids with a high specific heat capacity reach higher fractions of the Carnot COP. From this point of view, refrigerants with high molecular weights should be preferred. The trade-off is that the vapor pressure tends to be very low for these fluids and system size increases considerably. Thus, the low pressure refrigerants are best suited for large capacity machines.

For zeotropic mixtures, the suction to liquid line heat exchanger has an additional effect as shown in the $\ln(P)$-h diagram of Figure 4.2.4. By reducing the enthalpy at the evaporator inlet, the temperature of the entering mixture is reduced because of the influence of the glide. T_{5a} is lower than T_5, assuming that the suction pressure level did not change because of the heat exchanger. This decrease in temperature at the evaporator inlet increases the evaporator capacity because the temperature difference that drives the heat transfer is increased.

Alternatively, the suction pressure can be increased slightly and the penalty of increased compressor work somewhat reduced. Depending on the size of the glide and its linearity, this effect can be significant (steep glide close to the liquid line) or nonexistent (flat glide close to the liquid line). See Figures 4.1.3 and 4.1.4 with examples in Chapter 5 where experimental results are discussed.

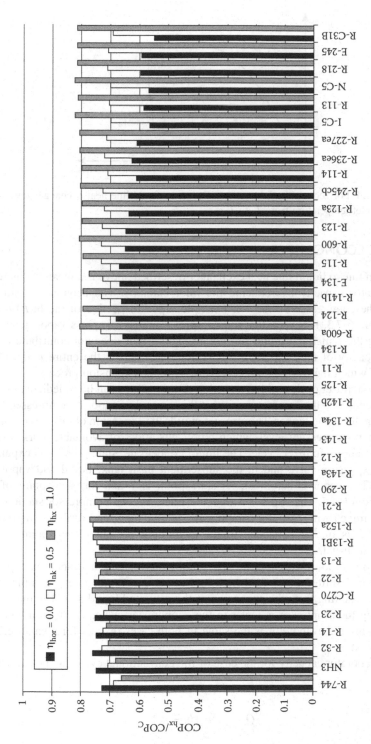

FIGURE 4.2.3 Plot of COP improvement using suction to liquid line heat exchanger (Domanski, 1995).

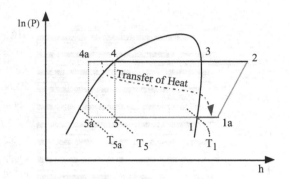

FIGURE 4.2.4 For zeotropic mixtures, the suction to liquid line heat exchanger lowers the evaporator inlet temperature from T_5 to T_{5a}.

4.3 THE ECONOMIZER

The suction line to liquid line heat exchanger described above is a means to increase the evaporator capacity by reducing the inlet enthalpy. The economizer aims at reducing the compressor power. For example, upon inspection of the $\ln(P)$-h diagram, Figure 3.7.3, it is seen that the compressor must compress vapor generated for cooling the remaining liquid refrigerant. This vapor does not contribute to the cooling capacity of the evaporator but is generated across the entire temperature range over which the refrigerant is cooled during the expansion process.

The economizer cycle extracts a portion of this vapor at an intermediate pressure level and sends it to the compressor so that this portion of vapor is compressed from that pressure level and higher. This process avoids the expansion of this vapor portion to the evaporator pressure level, which represents a loss of avoidable work. Figure 4.3.1 shows an economizer cycle. The liquid condensate, state point 3, is expanded in a first expansion valve and in the subsequent flash tank, liquid and vapor are separated. The vapor is saturated; its state point is 7 and the state point of the saturated liquid is 5. The vapor 7 is directed into a second compressor suction port where it is mixed with partially compressed suction vapor and further compressed. The liquid 5 is further expanded to the evaporator pressure level, point 6, and evaporated as usual.

Figure 4.3.1 shows what appears to be an obvious increase in evaporator capacity. State point 6 is shifted much further toward lower evaporator inlet enthalpies than at point 6a, resulting from the expansion of the saturated liquid at 3. The mass flow rate entering the evaporator is lower in the economizer cycle. The total amount of liquid entering the evaporator is the same in both cases (except for a small effect) as shown next.

The evaporator capacity for the case of the conventional cycle can be calculated as

$$\dot{Q}_{evap} = \dot{m}_3(h_1 - h_3) \qquad (4.2)$$

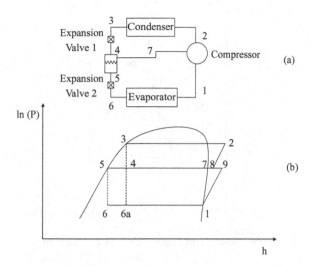

FIGURE 4.3.1 (a) State points superimposed on a ln(P)-h diagram. (b) Schematic of a vapor compression cycle with economizer.

and for the economizer cycle as

$$\dot{Q}_{evap,Eco} = \dot{m}_3 (1 - x_4)(h_1 - h_5) \tag{4.3}$$

where x_4 is the fraction of the total mass flow rate that is extracted on the intermediate pressure level.

The difference in evaporator capacity can be calculated as

$$\Delta\dot{Q} = \dot{Q}_{evap,Eco} - \dot{Q}_{evap} \tag{4.4}$$

When we calculate $\Delta\dot{Q}$, the following expression is obtained:

$$\Delta\dot{Q} = \dot{m}_3 [(1 - x_4)(h_1 - h_5) - (h_1 - h_3)] \tag{4.5}$$

where

x_4 is the vapor quality at point 4

x_4 can be expressed in terms of enthalpy as Equation 4.6

$$x_4 = (h_3 - h_5)/(h_7 - h_5) \tag{4.6}$$

After using Equation 4.6 for x_4, the following final expression Equation 4.7 is obtained:

$$\Delta\dot{Q} = \dot{m}_3 (h_3 - h_5) \frac{h_7 - h_1}{h_7 - h_5} \tag{4.7}$$

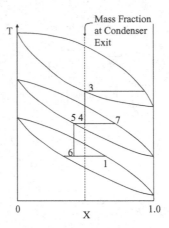

FIGURE 4.3.2 State points according to Figure 4.3.1 on *T-x* diagram.

Thus, the change in evaporator capacity because of the economizer feature is shown to be proportional to the reduction in evaporator inlet enthalpy $(h_3 - h_5)$ times a correction term that is based on the difference in vapor enthalpies of the flash tank exit (h_7) and the evaporator outlet (h_1). If this difference is zero, there is no benefit in the evaporator capacity derived from the economizer cycle as expected. This means both enthalpies are the same and therefore the saturated vapor line in the $\ln(P)$-h diagram is vertical in this region. Depending on whether h_7 is larger or smaller than h_1, the change in ΔQ can be positive or negative, respectively.

While the economizer cycle might or might not be of benefit to the evaporator capacity, it certainly has a benefit in terms of reduced compressor work for two reasons. First, the compressor work in the lower compression stage is reduced because the mass flow rate in the compressor is reduced to the amount of $(1 - x_4)$. Second, as can be seen on the $\ln(P)$-h diagram of Figure 4.3.1, the second stage of the compression process operates with a lower suction temperature than would be the case without the economizer.

When zeotropic refrigerant mixtures are used, the impact of the economizer cycle is more complex. The vapor leaving the flash tank contains predominantly the lower boiling component, while the liquid contains the higher boiling component. This is shown in the *T-x* diagram of Figure 4.3.2 with the state points the same as in Figure 4.3.1. Because of the separation in the flash tank, the mass fraction in the evaporator is lower. For the same evaporation pressure, the coldest temperature in the evaporator is now raised because of the composition shift. This effect is opposite the one described above for suction heat exchangers.

It should be noted that because of the composition shift, the economizer cycle cannot be correctly displayed on the $\ln(P)$-h diagram for a mixture. These diagrams are only valid for a given mass fraction and any change in mass fraction cannot be meaningfully displayed. This would require a switch to the enthalpy-mass fraction diagram. However, this diagram is not suitable for showing the superheated vapor region and therefore the compression process cannot be shown.

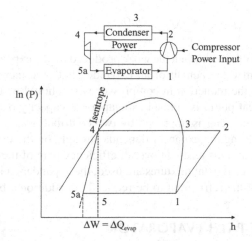

FIGURE 4.4.1 Vapor compression cycle with expander to recover expansion work.

To conduct performance calculations for a zeotropic mixture cycle with an economizer using just one $\ln(P)$-h diagram will lead to evaporator pressures that are slightly too high and enthalpy values that are not quite correct. The larger the difference in the boiling points of the two mixture constituents becomes, the higher the actual value of these deviations.

4.4 THE EXPANDER

The expansion process reduces the pressure of the refrigerant from the condenser level to the evaporator level in a process of constant enthalpy and is highly irreversible. As an alternative, a work-producing expansion device can be employed instead. The power output of this device can then be supplied to the compressor to reduce compressor power input. Figure 4.4.1 shows a schematic of this case. State point 5 represents the outlet condition after the conventional expansion process and state point 5a represents the post-isentropic expansion process. In case of the isentropic expansion, there is less refrigerant vapor produced in order to cool the fluid to the evaporator temperature level because a portion of the energy content is removed in the form of work. However, the saturation line is usually not reached. The isentropes in the two-phase region are steeper than the saturated liquid line in a $\ln(P)$-h diagram.

An inspection of Figure 4.4.1 reveals that the expander produces a double benefit. As mentioned above, it produces work that reduces the power requirement of the compressor. The amount of work produced by the expander can be calculated as

$$\Delta \dot{W} = \dot{m}(h_5 - h_{5a}) \tag{4.8}$$

This same amount of energy, since it is extracted from the refrigerant stream, also represents an increase of evaporator cooling capacity as shown in Figure 4.4.1. The new evaporator capacity is now calculated as

$$\dot{Q}_{evap} = \dot{m}(h_1 - h_{5a}) \tag{4.9}$$

This benefit is available even when the work produced by the expander is not utilized. However, to maximize the benefit of an expander, both the increase in evaporator capacity as well as the reduction in compressor work should be utilized. Since the efficiency of the heat pump is defined as evaporator capacity over net compressor power input, this parameter is improved by two contributions.

The benefit of using the expander depends strongly on the overall efficiency of the device. Current designs reach an overall efficiency gain of the heat pump on the order of 5%. When refrigerant mixtures are used, the expander causes the evaporator inlet temperature of the refrigerant to be reduced, an additional benefit.

4.5 THE THREE-PATH EVAPORATOR

The three-path evaporator offers a unique opportunity for zeotropic mixtures not available for pure fluids. Its effectiveness depends strongly on the presence of a large glide. Figure 4.5.1 demonstrates how the concept works. The first path in the evaporator is for the chilled fluid, here assumed to be water. The second path is for the evaporating refrigerant mixture and both paths are best arranged in counter flow. The third path is for the high pressure liquid refrigerant that leaves the condenser (or, as an alternative, exits the suction line to liquid line heat exchange, not shown in Figure 4.5.1). With this internal heat exchange process, the evaporating refrigerant mixture cools the water as well as the liquid refrigerant. Because of the second heat source, the cooling capacity of the evaporating refrigerant seems to be reduced. However, at the same time the enthalpy of the liquid refrigerant is reduced by the same amount; in terms of cooling capacity, there is no net effect. Nevertheless, when mixtures are used, this measure leads to a reduced temperature at the evaporator inlet. The glide is increased as well if the pressure level is constant, demonstrated in Figure 4.5.2. The net effect is very similar to increasing the degree of subcooling at the condenser outlet, as shown in Figure 4.1.2. However, in that case the evaporator

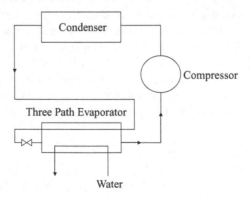

FIGURE 4.5.1 Three-path evaporator.

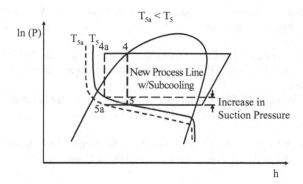

FIGURE 4.5.2 The three-path evaporator lowers the evaporator inlet temperature. Alternatively, the suction pressure can be raised.

capacity is increased and the inlet temperature lowered, also beneficial for pure fluids. In the case of the three-path heat exchanger, no benefit is derived for pure fluids because there is no net gain in evaporator capacity while there is a reduction in the evaporator inlet temperature of the two-phase refrigerant stream for mixtures. As discussed above, this benefit again can be translated into a somewhat higher compressor suction pressure when the conditions in the evaporator allow. This condition is fulfilled when the evaporator outlet temperature after the increase in suction pressure is still lower than the incoming water temperature.

The main difference between improved subcooling at the condenser outlet and the three-path evaporator is the following: When the liquid refrigerant is subcooled in the condenser, the heat of subcooling is rejected to the exterior where it leaves the cycle. With the three-path evaporator, the heat of subcooling remains within the cycle. Therefore, no capacity benefit is achieved but a secondary benefit (reduced evaporator inlet temperature) is acquired in the case of zeotropic mixtures. The benefit of the three-path evaporator is available in addition to that of the suction to the liquid line exchanger and there is no operational penalty associated with it.

Ideally, and especially for mixtures with high specific heat capacities, a designer should consider employing all three measures to the maximum beneficial extent: subcooling, suction line to liquid line exchange and the three-path evaporator configuration.

REFERENCES

Domanski, P.A., 1995, "Theoretical Evaluation of the Vapor Compression Cycle With a Liquid-Line/Suction-Line Heat Exchanger, Economizer, and Ejector," NIST Report, NISTIR 5606.

Radermacher, R., 1988, "Vapor Compression Heat Pump with Non-azeotropic Refrigerant Mixture and Absorber Desorber Heat Exchange," U.S. Patent No. 4,724,679.

Radermacher, R. and D.S. Jung, 1993, U.S. Patent No. 5,207,077.

Vakil, H.B., "Means and Methods for the Recovery of Expansion Work in Vapor Compression Cycle; Subcooling in Tube Parallel to Evaporator," U.S. Patent No. 4,304,099.

Vakil, H.B., "Vapor Compression Cycle Device with Multicomponent Working Fluid Mixture; Receiver Behind Condenser and in Between the Two Evaporators," U.S. Patent No. 4,283,919.

Vakil, H.B., "Vapor Compression Cycle Device with Multicomponent Working Fluid Mixture and Method; Receivers Behind Condenser and Evaporators," U.S. Patent No. 4,217,760.

Vakil, H.B. and J.W. Flock, "Vapor Compression Cycle Device with Multicomponent Working Fluid Mixture and Method; Pseudo Counter Flow Heat Exchanger," U.S. Patent No. 4,281,890.

Vakil, H.B., "Vapor Compression Cycle Device with Multicomponent Work Fluid Mixture; Receiver and Separate Separation Device Behind Cond. Accumulator," U.S. Patent No. 4,179,898.

5 Experimental Performance Measurements

Section 3.6 outlines the Lorenz cycle and its potential for improved efficiency under properly selected operating conditions. These occur in a heat pump application when there is a large and approximately equal temperature glide in condenser and evaporator and a temperature lift that is as small as possible as compared to the glide. One such application is air-conditioning, where for most operating hours the air temperatures inside and outside a building are essentially equal. An ARI Standard 210/240 (1994) specifies the return air temperature in a building as 26.7°C and the outdoor temperature as 27.8°C. In order to ensure adequate dehumidification, the air supplied to the inhabited space must be cooled to about 12.8°C. Thus, the temperature glide in the evaporator is 14 K. Accordingly, the temperature glide in the condenser should be at least 14 K or even more as described in Section 3.8; here it is chosen to be 19 K. Figures 3.8.1 and 3.8.2 show a schematic of the temperature conditions involved.

Although the air stream temperatures reflect almost the ideal operating conditions (it would be better if both air streams would start from the same temperature), the situation is not quite so ideal for the refrigerant. Since there is always a final approach temperature for a realistic heat exchanger, the lowest temperature of the refrigerant leaving the condenser and the warmest leaving the evaporator are more than 1 K apart. To demonstrate this performance potential, a key experiment is described in this chapter. It will be discussed here in considerable detail because it highlights opportunities and challenges of working fluid mixtures very clearly. The experimental setup is also a very good example for how reliable heat pump tests can be conducted with the necessary checks and redundancy to ensure reliable test data. In this regard, this chapter introduces experimental evaluation of refrigerant mixtures in a laboratory breadboard and actual heat pump systems. Some of experimental results of refrigerant mixtures using these test facilities are then explained (Vineyard, 1988; Vineyard and Conklin, 1991; Conklin and Vineyard, 1991; Vineyard, Conklin, and Brown, 1993; Kauffeld et al., 1990; Pannock and Didion, 1991).

5.1 LABORATORY BREADBOARD TESTS

In this section, three sets of experimental investigations are introduced on the performance of mixtures using laboratory breadboard heat pumps conducted at NIST (Kauffeld et al., 1990; Pannock and Didion, 1991) and ORNL (Vineyard, 1988;

Vineyard and Conklin, 1991; Conklin and Vineyard, 1991; Vineyard, Conklin, and Brown, 1993).

5.1.1 NIST Breadboard Heat Pump I

The following text is based on excerpts of the above mentioned report. The full text is available as NISTIR 90-4290.

5.1.1.1 Test Facility

The breadboard heat pump I is shown schematically in Figure 5.1.1. Water was selected as a heat source- and sink-fluid because capacity measurements are simple and because counterflow and intracycle heat exchange can be achieved. A constant evaporator heat flux was desired to compare the different refrigerants (McLinden and Radermacher, 1987). Because the refrigerants to be examined had differing properties, achieving this goal required use of a variable speed compressor. An open two piston reciprocating compressor was selected together with a variable speed motor.

Figure 5.1.2 shows a side view of the setup. The condenser heat exchangers are coiled aluminum extrusions, as shown in Figure 5.1.3. The evaporator heat exchangers (Figure 5.1.4) were chosen to allow for temperature measurements all along the flow channel. Two 6 m long extruded copper tubes were selected for the evaporator heat exchanger. The heat exchanger tubes consisted of only two passages, a center path with "microfins" on the walls and a set of "rectangular" channels around this inner tube. The fins and channels are spiraled, changing top to bottom every 150 mm. A third passage was added on the outside of the evaporator to obtain a heat exchanger for a third fluid, thus allowing for heat exchange between the evaporator refrigerant and the condensed liquid refrigerant throughout the length of the evaporator. The two heat exchanger tubes are connected in series, resulting in a total evaporator heat exchanger length of 12 m.

Thermocouples were soldered to the outside of the evaporator heat exchangers every 0.6 m, thereby allowing measurement of the temperature profile in the outermost channel. For the rectangular channels, measurement was limited to the crossover between 3 m sections of the 12 m heat exchanger. This resulted in five thermocouples for these passages. For the center passage, thermocouples are located every 0.6 m in a sealed 3.2 mm thin wall, stainless steel tube. These thermowells are mounted in the center of the innermost channel to measure the in-stream temperature of the flowing fluid. Additional measurement devices were several other thermocouples, thermopiles and pressure transducers located at critical points in the cycle.

Another important feature of the breadboard heat pump is a liquid/vapor-separating accumulator in front of the compressor serving as a refrigerant storage vessel as well (Figure 5.1.5).

5.1.1.2 Capacity Measurements

The capacity of both heat exchangers was measured using redundant methods for verification and the compressor power was measured using a torque meter and

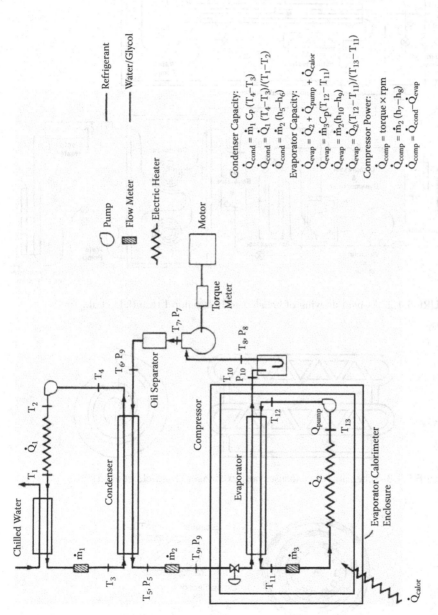

FIGURE 5.1.1 Schematic of breadboard heat pump I (Kauffeld et al., 1990).

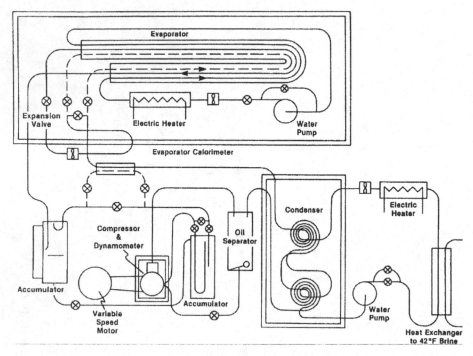

FIGURE 5.1.2 As-built drawing of breadboard heat pump I (Kauffeld et al., 1990).

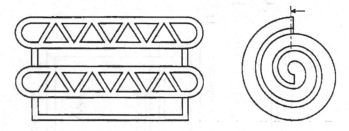

FIGURE 5.1.3 Schematic of condenser heat exchanger (Kauffeld et al., 1990).

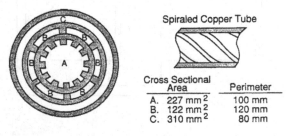

Cross Sectional Area	Perimeter
A. 227 mm^2	100 mm
B. 122 mm^2	120 mm
C. 310 mm^2	80 mm

FIGURE 5.1.4 Schematic of evaporator extruded heat exchanger (Kauffeld et al., 1990).

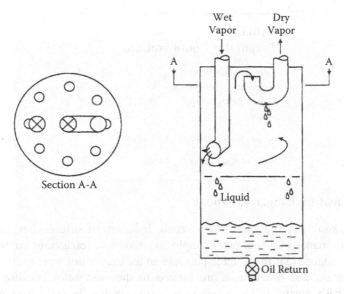

FIGURE 5.1.5 Schematic of accumulator (Kauffeld et al., 1990).

tachometer. All the different methods employed throughout the test program agreed closely (within 6%), confirming the validity of the test results.

5.1.1.3 Test Procedure

Two zeotropic binary refrigerant mixtures R22/R114 and R13/R12 were tested in the breadboard heat pump. Tests were conducted at the same capacity for each mixture to achieve equal average heat flux per unit area of the heat exchangers for all tests as one of the major criteria for comparability of refrigerants. One exception was pure R114, which would have required an excessively high compressor speed to reach the desired capacity. The capacity was chosen to be 4.1 kW, obtained with pure R22 at 500 RPM compressor speed. Heat sink and heat source entry and exit water temperatures were kept constant for all tests as the second major criterion for comparability (McLinden and Radermacher, 1987). The temperatures were 28 and 47°C entering and leaving the condenser, and 27 and 13°C entering and leaving the evaporator. The expansion device was set simultaneously to give marginal subcooling at the condenser outlet and to provide evaporation throughout the entire evaporator. The subcooling was determined by the disappearance of vapor bubbles in a sight glass in the liquid line upstream of the expansion valve. The condition of flooded evaporator was determined by the presence of liquid droplets leaving the evaporator in a sight glass in this pipe and by the presence of liquid in the accumulator.

The mixture mass fractions examined cover the whole range from 0 to 100% for the R22/R114 test series, whereas for the second refrigerant mixture (R13/R12) mass fractions were limited to less than 45% of R13 by weight because of excessive pressures in the condenser that exist with more R13.

TABLE 5.1
Evaporator Configurations

Configuration No.	1	2	3	4
Evaporating Refrigerant	C	B	A	A
Liquid Refrigerant	B	C	B	C
Water	A	A	C	B

A: inside, B: rectangles, C: outside

Source: Kauffeld et al., 1990.

5.1.1.4 Heat Exchanger Variations

A major task of this project was to analyze the influence of different heat exchangers on the performance of a heat pump employing zeotropic refrigerant mixtures. Four different configurations of the three passages in the evaporator were tested. Detailed instrumentation and observation are limited to the evaporator because this heat exchanger has a greater effect on system performance than the condenser. According to the nomenclature of Figure 5.1.4, the four evaporator configurations are shown in Table 5.1.

Configurations 2 and 4 were expected to perform better than 1 and 3 since the evaporating refrigerant and water (which constituted the majority of the transferred heat) were in the channels that had enhanced surfaces. Furthermore, these channels were adjacent, limiting the losses due to heat passing through the intermediate passage.

The performance of configurations 1 and 3 is strongly influenced by the surface characteristic of the outermost channel. The smooth copper surface provides a poor heat exchange area, resulting in large temperature differences between the water and the refrigerant for these two configurations. The relative heat exchanger performance is shown in the temperature plots for the four configurations for pure R22 in Figure 5.1.6. It is observed that R22 uses the full length of the heat exchanger for configurations 1 and 3 but only about one half to three quarters of the heat exchanger for configurations 2 and 4. Figure 5.1.7 shows the evaporator temperature profiles of R22/R114. As shown, the heat exchangers for configurations 2 and 4 have smaller approach temperatures than that of R22 due to the temperature glide of R22/R114. Figures 5.1.8 and 5.1.9 show the performance of R22/R114.

For the mixture of R22/R114, a maximum increase in coefficient of performance (COP) of 32% was found at a mass fraction of 65% of R22. At lower R22 mass fraction, the capacity dropped off rapidly and the COP of R114 was determined to be about 40% lower than that of R22. However, this large decline is a reflection of the fact that the compressor is not designed for such a low-pressure refrigerant and valve pressure losses become dominant. For R22/R114 there is no significant difference in performance whether or not the suction-line heat exchanger is used (Figure 5.1.8). The best performance was achieved with evaporator configuration 4. Configuration 3 performed worst but still showed an increase in performance of 5 to 10%.

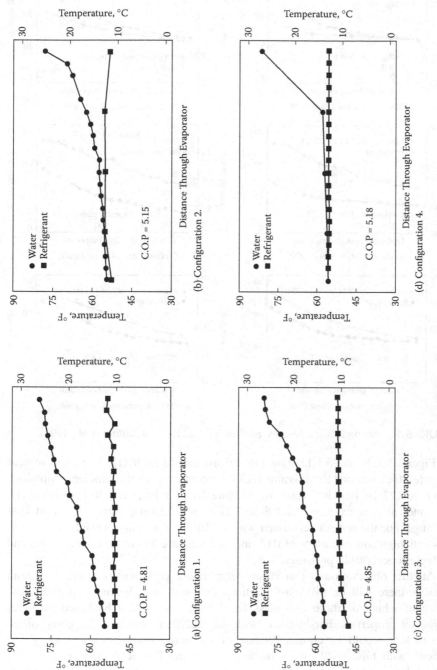

FIGURE 5.1.6 Evaporator temperature profiles for R22 (Kauffeld et al., 1990).

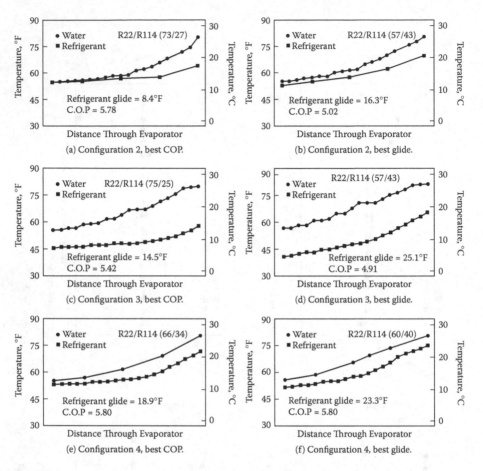

FIGURE 5.1.7 Evaporator temperature profiles for R22/R114 (Kauffeld et al., 1990).

Figures 5.1.10 and 5.1.11 show the performance of R13/R12. The improvement in cycle efficiency for the mixture R13/R12 compared to the efficiency obtained with pure R22 and pure R12 is shown in Figure 5.1.11 to be 16 and 40%, respectively, with internal heat exchange and 8 and 27%, respectively, without internal heat exchange. For the test set performed with R13/R12, it was not possible to perform tests with high concentrations of R13 since the pressure would have been far beyond the highest acceptable pressures.

Another observation is that at any temperature glide less than the maximum possible, there will be two mass fractions that will give the required glide. The choice of which of these two mass fractions to use would be based on other refrigerant properties. For instance, with the R22/R114 mixture, the composition high in R22 would be chosen because of the better compressor efficiency and capacity with higher R22 mass fractions. With the R13/R12 mixture, the mass fraction low in R13 would be chosen to avoid the high discharge pressures with high mass fractions of R13.

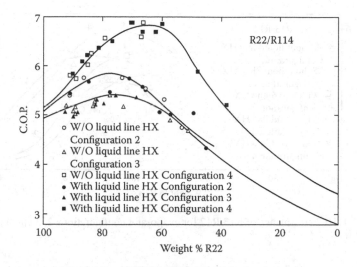

FIGURE 5.1.8 Performance of R22/R114 for evaporator configurations 2, 3 and 4 (Kauffeld et al., 1990).

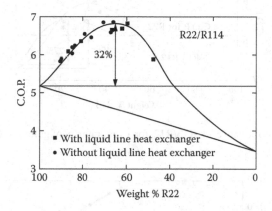

FIGURE 5.1.9 Performance of R22/R114 for evaporator configuration 4 (Kauffeld et al., 1990).

The capacity per compressor revolution for both mixtures in evaporator configuration 4 is shown in Figure 5.1.12. Compared to pure R22, the R22/R114 mixture (65/35 wt.%) that showed 32% efficiency improvement (Figure 5.1.10) resulted in a capacity loss of 8%. The R13/R12 mixture (34/66 wt.%) that showed a 16% efficiency improvement over pure R22 (Figure 5.1.11) resulted in a capacity increase of 21%.

5.1.1.5 Linearity of the Enthalpy vs. Temperature Relationship

An interesting feature of temperature plots is that the location of the pinch points varies for different refrigerant mixtures. The *pinch point*, the closest temperature

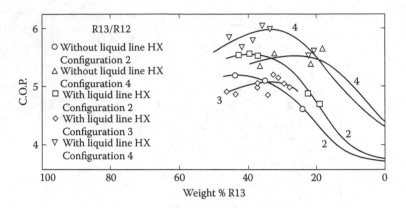

FIGURE 5.1.10 Performance of R13/R12 for evaporator configurations 2, 3 and 4 (Kauffeld et al., 1990).

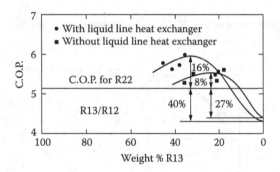

FIGURE 5.1.11 Performance of R13/R12 for evaporator configuration 4 (Kauffeld et al., 1990).

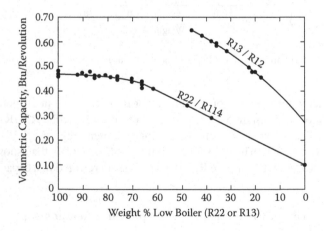

FIGURE 5.1.12 Volumetric capacity of R22/R114 and R13/R12 for evaporator configuration 4 (Kauffeld et al., 1990).

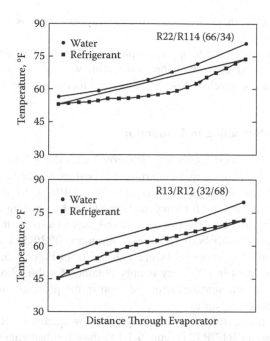

FIGURE 5.1.13 Comparison of evaporator temperature profiles for R22/R114 and R13/R12 (Kauffeld et al., 1990).

approach between the refrigerant and the heat exchange fluid, is at the two ends of the heat exchanger for the R22/R114 mixture, whereas for the R13/R12 mixture the pinch point is in the center of the heat exchanger. This is shown in Figure 5.1.13, where a line connects the points of entry and exit to emphasize this nonlinearity.

The nonlinear shape of the temperature profile during evaporation results from the differing properties of the pure refrigerants in combination with mixture effects. The dominant effect in determining the shape of the profiles appears to be the relative amounts of the more and less volatile components. With the R22/R114 mixture of 60% R22, the bulk of heat necessary to vaporize the mixture is required at temperatures close to the bubble point temperature (i.e., at temperatures closer to the boiling point of the more volatile R22). As a result, the temperature versus enthalpy profile is concave. With the R13/R12 mixture, the mass fraction of the more volatile R13 is only 20% and thus the bulk of the heat of vaporization is concentrated at the higher temperatures, resulting in a convex profile.

A secondary effect is the different heats of vaporization for the two components. This is most pronounced with R22/R114, where the heat of vaporization of R22 is substantially higher than that of R114. Because the first bubbles of vapor generated are enriched in R22, the higher heat of vaporization for this refrigerant further contributes to the concave shape. This behavior is important for the prediction of system performance by the use of simple computer programs which assume a linear temperature glide throughout the heat exchanger, since the actual glide may deviate quite far from linearity. These errors are greatest with efficient heat exchangers where the temperature differences are small, the application seen as most favorable to mixtures.

The convex shape of the temperature profile of R13/R12 tends to be amplified by a pressure drop as present in the evaporator, whereas the concave shape of R22/R114 tends to be leveled. Another aspect of this nonlinearity is discussed in the next section.

5.1.1.6 Liquid Precooling in Evaporator

The final part of this investigation was concerned with the validity of the proposed improvement obtained by using an intermediate heat exchange between the evaporating refrigerant and the condensed liquid (Vakil, 1981). Liquid precooling in evaporator (LPE) improves efficiency with mixtures by shifting a portion of the evaporator capacity to a lower vapor quality and thus to a lower temperature. Tests have been conducted with both refrigerant mixtures R22/R114 and R13/R12 to evaluate this theoretical proposal. As one can see from Figures 5.1.10 and 5.1.11, noticeable improvement in efficiency is only obtained for the mixture of R13/R12. The reasons for this different behavior are seen in the pressure-enthalpy diagrams for the two refrigerant mixtures.

First, the two-phase isotherms are nearly flat at low quality for R22/R114 (Figure 5.1.14) while those for R13/R12 (Figure 5.1.15) show a substantial slope throughout the two-phase region. Because of this flatness, the heat exchanger has little value

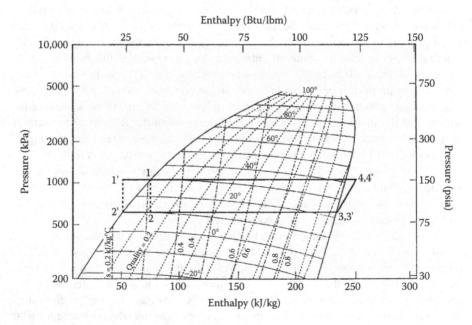

FIGURE 5.1.14 Effect of suction line heat exchanger for R22/R114 (Kauffeld et al., 1990).

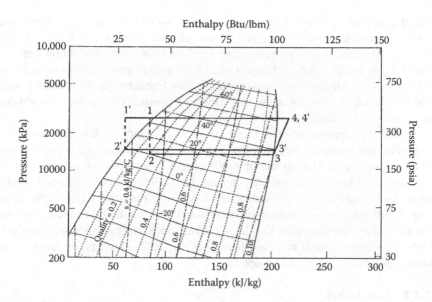

FIGURE 5.1.15 Effect of suction line heat exchanger for R13/R12 (Kauffeld et al., 1990).

for R22/R114 for the same reasons that cause it to have little value for pure refrigerants with their truly horizontal isotherms. Referring to Figures 5.1.14 and 5.1.15, the shift of the evaporator entering condition from point 2 to 2 can be seen to result in a higher evaporator pressure (which improves efficiency) at the same average evaporator temperature for R13/R12, while the shift from point 2 to 2 has no effect on evaporator pressure for R22/R114. Second, the shape of the two-phase dome results in flashing more of the R13/R12 mixture between the approximate operating points (30°C condensing, 10°C evaporating) at which the tests were run, indicating more potential for reducing entering vapor quality.

5.1.1.7 Discussion

By using zeotropic refrigerant mixtures with their changing temperatures at constant pressures in the two-phase region, one can obtain a temperature glide on the refrigerant side in a conventional heat exchanger. A zeotropic mixture at the right mass fraction can match the temperature glide of the heat exchange fluid in a counter flow heat exchanger.

However, special care must be taken in the heat exchanger design to suit the requirements of the zeotropic mixtures. This is one of the reasons why several earlier experimental studies failed to show the projected improvements. Because pure refrigerants quickly encounter a pinch point as the heat exchanger area is increased, mixtures can take much better advantage of increased areas. For instance, with the most effective heat exchanger configuration, the pure refrigerant could only effectively use a quarter of the heat exchanger area because of its pinch point and would be expected to produce substantially the same efficiency with its heat exchangers reduced in size to a quarter of that employed. It must be emphasized, however, that all the comparisons between pure and mixed refrigerants presented in this chapter are for equal areas. A compromise must be found for each application between increased cost for the heat exchangers and obtainable improvements in efficiency, resulting in cost savings when operating the system. Furthermore, the heat exchanger should not result in a high pressure drop since the pressure drop affects the temperature glide of a mixture.

Finally, the proposed improvements in cycle efficiency by the introduction of an internal heat exchange between the evaporating refrigerant and the condensed liquid is proven valid. The amount of improvement is dependent on the shape of the vapor dome and temperature profiles and can be small for some mixtures. It is also shown that the introduction of such a heat exchanger does not significantly affect the cycle efficiency of single component systems. The potential for cycle improvement is present with mixtures. Employing zeotropic refrigerant mixtures results in an efficiency improvement of as much as 32% over R22 when heat exchangers and temperature glides are selected properly.

5.1.1.8 Conclusion

- For all heat exchanger configurations for which a full range of mass fractions was tested, the optimum mass fraction of each mixture performed better than pure R22.
- The best efficiency measured with the mixture R22/R114 (at a mass fraction of 65% R22) was approximately 32% higher than the best efficiency measured with R22.
- The best efficiency measured with the mixture R13/R12 (at a mass fraction of 35% R13) was approximately 16% better than the best efficiency measured with R22.
- The better the heat exchanger effectiveness, the greater the improvement the mixtures showed over pure R22.
- In the evaporator configurations having poorer heat exchanger effectiveness, the gliding temperature effect of mixtures was less important to cycle efficiency than other refrigerant properties.
- Pressure drop in the evaporator reduced the temperature glide and hence reduced the cycle efficiency improvement by mixtures.
- Nonlinearity of temperature with enthalpy was observed for the two mixtures tested.
- Intracycle heat exchange between the condensed liquid and evaporating refrigerant streams benefited the R13/R12 mixture but not the R22/R114 mixture.
- Intracycle heat exchange between the condensed liquid and evaporating refrigerant streams is beneficial for those mixtures showing a substantial loss of potential temperature glide as a result of flashing through the expansion device.
- Efficiency improvement by using mixtures does not imply loss of capacity as a trade-off.

5.1.2 NIST Breadboard Heat Pump II

The following text is based on excerpts of an NIST internal report. The full text is available in NISTIR 91-4748.

The laboratory breadboard heat pump II was designed for a 3.5 kW capacity. Other design criteria were:

- Counterflow heat exchanger configuration for condenser, evaporator and suction line heat exchanger (SLHX).
- Liquid heat transfer fluid.
- Variable speed compressor.
- System can be operated with or without SLHX.
- Accessibility of refrigerant and heat transfer fluid in order to obtain condenser and evaporator temperature profiles.

5.1.2.1 Test Facility

Figure 5.1.16 shows a schematic of the laboratory breadboard heat pump II. The refrigerant side consists of the compressor, condenser, SLHX, expansion device and evaporator. To minimize the charge, the system components were limited to the basic components without any large volume auxiliary equipment such as an oil separator or an accumulator. The compressor was two cylinder reciprocating compressor. All heat exchangers used in this breadboard heat pump II (evaporator, condenser and SLHX) were in a counterflow heat exchange configuration: the refrigerant flows in the inner tube and a water-ethylene glycol mixture flows in the annulus. The inner tube and annulus side were equipped with turbulators and spine fins in order to increase the heat transfer coefficient. This heat transfer enhancement resulted in large flow pressure drops but had other benefits such as the compactness of the system design and reduced refrigerant charge. The SLHX was installed between the liquid line after the condenser outlet and compressor suction line with a bypass line. The expansion valve was a metering valve controlled manually.

5.1.2.2 Test Results of Test Apparatus II

For all tests, two cycle parameters were controlled such that the degree of subcooling at the condenser outlet and degree of superheating at the evaporator outlet were positive in order to get comparable test results. The degree of subcooling was especially limited as much as possible (not greater than 2 K) and the superheating region was limited within 5% of the evaporator length. These conditions were satisfied by adjusting the refrigerant charge and expansion valve opening. The capacity was maintained for all tests by adjusting the compressor speed from 1500 RPM to 1620 RPM. A refrigerant sample was taken from the discharge line during the steady state phase of the test and then evaluated with the gas chromatograph to measure the mass fraction of the mixtures.

Using this test apparatus II, R32/R134a and R32/R152a mixtures at various mass fractions of R32 were conducted at various test conditions defined in Table 5.2.

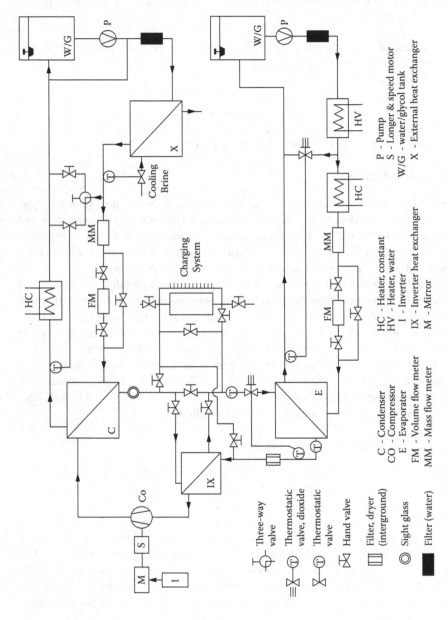

FIGURE 5.1.16 Schematic of breadboard heat pump II (Pannock and Didion, 1991).

TABLE 5.2
Test Conditions for Heat Transfer Fluid

Test	Evaporator Inlet (°C)	Evaporator Outlet (°C)	Condenser Inlet (°C)	Condenser Outlet (°C)
High temperature cooling (A)	26.7	14.4	35.0	43.2
Low temperature cooling (B)	26.7	14.4	27.8	43.2
High temperature heating (47S)	8.3	14.4	21.1	43.2
Low temperature heating (17L)	8.3	14.4	21.1	43.2

Source: ASHRAE, 1995.

High Temperature Cooling Test (Test A)

The cooling capacity was kept constant at 3.5 kW for these tests. The compressor speed of R32/R134a and R32/R152a was the same as that of R22 when the mass fractions of R32/R134a and R32/R152a are at 24/76 wt.% and at about 31/69 wt.%, respectively. Figure 5.1.17 shows the COP improvement over R22 for Test A. The COP improvement increases steadily when the R32 mass fraction increases for both mixtures. For R32 mass fraction below 22% in the R32/R134a mixture, the COP is lower than that of R22. The maximum COP improvement reached at 19% over R22 at 40% of R32 mass fraction. For R32/R152a, the COP is higher than that of R32/R134a for mass fractions lower than about 37%. The maximum COP increase was 17 to 18% over that of R22 at the same capacity. The COP was 14 and 4% higher for R32/R152a and R32/R134a, respectively, at the mass fraction having the same speed and capacity with R22.

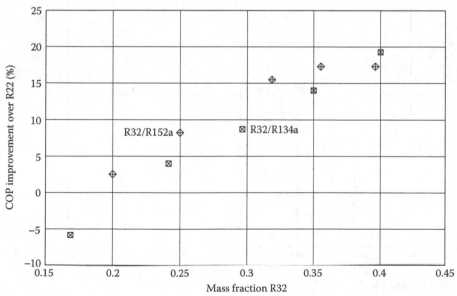

Test condition 1A

FIGURE 5.1.17 Cooling COP for Test A (Pannock and Didion, 1991).

The R32 mass fraction of the mixture affects the suction and discharge pressure. Lowering R32 mass fraction can reduce the operating pressures but it also reduces volumetric capacities. R32/R152a mixtures showed lower pressures than R32/R134a mixtures at the same R32 mass fraction. At a mass fraction of 28 and 36%, the suction pressures of R32/R134a and R32/R152a are equal to that of R22. A similar trend to the suction pressures was observed for the discharge pressures. The discharge pressures of R32/R134a and R32/R152a vary from −12 to +8% and from −24 to −6% with respect to R22 in the measured mass fraction range, respectively.

For both mixtures, the suction temperature was constant (within ±1 K) for all tests but the discharge temperature was different. For both mixtures, the discharge temperature was lower than that of the R22. The R32/R152a mixtures showed a decrease of 3 K to 5 K with increasing R32 mass fraction and the R32/R134a mixture showed a decrease of 5 K to almost 10 K as compared to R22.

The operating pressures and temperatures of both mixtures are within acceptable ranges as expected from the test condition.

Low Temperature Cooling Test (Test B)

The cooling capacity was kept constant at 3.7 kW for these tests. Figure 5.1.18 shows the COP improvement over R22 for Test B. Figure 5.1.18 shows that the COP improvement increases steadily when the R32 mass fraction increases for both mixtures as similar to Test A case.

For R32 mass fraction below 17% in the R32/R152a mixture, the COP is lower than that of R22. The COP improvement was over 20% when the R32 mass fraction

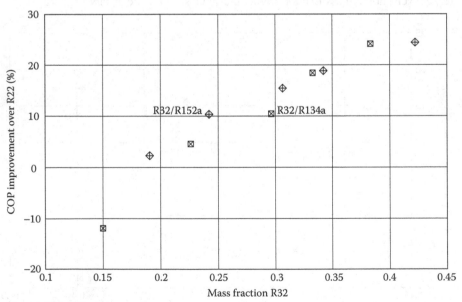

Test condition 1B

FIGURE 5.1.18 Cooling COP for Test B (Pannock and Didion, 1991).

was higher than 35%. For R32/R134a, the COP is lower than that of R32/R152a for R32 mass fractions lower than about 30%. The maximum COP increase was 22% over that of R22 at the same capacity at 38% R32 mass fraction.

At a mass fraction of 28 and 36%, the suction pressures of R32/R134a and R32/R152a are equal to that of R22. A similar trend to the suction pressures was observed for the discharge pressures. The discharge pressures of R32/R134a and R32/R152a vary from −12 to +8% and from −24 to −6% with respect to R22 in the measured mass fraction range, respectively.

Similar to Test A, the suction and discharge pressures were again within acceptable limits with respect to R22. The suction pressures of the R32/R134a mixture were 10 to 15% higher than those of the R32/R152a mixture, meaning about 100 kPa higher suction pressure. The discharge side pressure difference was about doubled to 200 kPa. The discharge pressures of R32/R134a were lower than those of R22 for R32 mass fraction of less than 30%. For R32/R152a, all measured discharge pressures were below that of R22.

For both mixtures, the suction temperature was constant (within ±1 K) for all tests except one data point but the discharge temperature was different. For both mixtures, the discharge temperature was lower than that of the R22. The R32/R152a mixtures showed a decrease of 3 to 5 K for 20 to 30% of R32 mass and the R32/R134a mixture showed a decrease of 10 K with more than 30% R32 mass fraction.

High Temperature Heating Test (Test 47S)

The heating capacity was kept constant at 3.1 kW for these tests. Figure 5.1.19 shows the COP improvement over R22 for Test 47S. Again, the COP improvement increases steadily when the R32 mass fraction increases for both mixtures as similar to the cooling mode. Different from the cooling mode, the heating COPs of the R32/R134a mixtures were higher than those of the R32/R152a mixtures (Figure 5.1.20). The COPs of both mixtures were higher than that of R22 when the R32 mass fraction was more than 20%. The R32/R152a mixture also shows a COP improvement against R22 when the R32 mass fraction was more than 20%. At the mixture compositions of same speed and same capacity, R32/R134a and R32/R152a showed 4 and 8% improvement, respectively.

Low Temperature Heating Test (Test 17L)

The heating capacity was kept constant at 2.0 kW for these tests. Figure 5.1.20 shows the COP improvement over R22 for Test 17L similar to the Test 47S case. The COP improvement increases steadily when the R32 mass fraction increases for both mixtures, as with the other cases. Similar to Test 47S, the heating COPs of the R32/R134a mixtures were 5 to 10% higher than those of the R32/R152a mixtures at the same R32 mass fraction (Figure 5.1.20). The COPs of both mixtures were lower than that of R22 for most of the R32 mass fractions investigated. Only the R32/R134a mixture showed a 5% COP improvement against R22 at 40% of R32 mass fraction. The same COP as R22 was reached at about 30% of R32 mass fractions for the R32/R134a mixtures and at above 40% of R32 mass fractions for the R32/R152a mixtures.

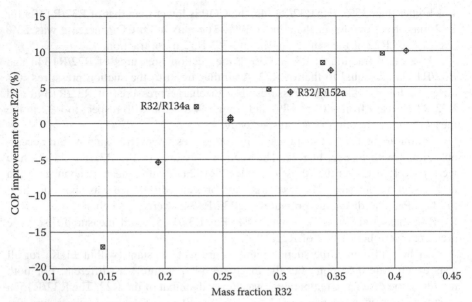

Test condition 1B

FIGURE 5.1.19 Heating COP for Test 47S (Pannock and Didion, 1991).

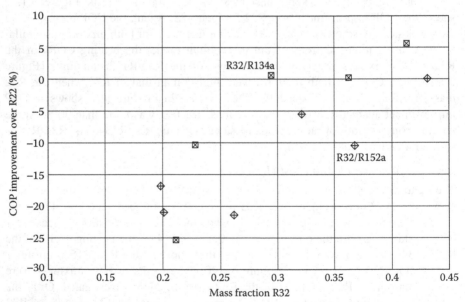

Test condition 10

FIGURE 5.1.20 Heating COP for Test 17L (Pannock and Didion, 1991).

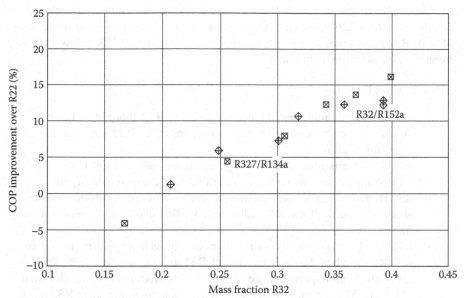

Test condition 1A-SLHX

FIGURE 5.1.21 Cooling COP for Test SLHX (Pannock and Didion, 1991).

Suction-Line Heat Exchange Test (Test SLHX)

The SLHX was tested at the same operating conditions as Test A (high temperature cooling). For a fair comparison, R22 was also tested with a SLHX and used as a reference. The cooling capacity was kept constant at 3.5 kW for these tests. Figure 5.1.21 shows the COP improvement over R22 for Test SLHX. The COP improvement increases steadily for both mixtures when the R32 mass fraction increases. For R32/R134a mixtures, the COP of R32/R134a matched the COP of R22 at 20% of R32 mass fraction and the COP was 17% higher than R22 at 40% of R32 mass fraction. For R32/R152a, the COP improvement was similar to R32/R134a except for the R32 mass fractions over 35%.

Overall, the SLHX provided a performance improvement, 4% in COP for R32/R152a and 5.8% in COP for R32/R134a. However, lower compressor speeds were used to run at the same capacity. This might contribute to the better compressor efficiency and COP improvement. If the compressor speed is the same as with R22, the COP improvement of the R32/R134a mixture and R32/R152a mixture was 3 and 11%, respectively, at the R32 mass fraction of the same capacity with R22. The use of the SLHX resulted in a lower refrigerant mass flow rate at the same capacity since the latent heat of evaporation increases by using the SLHX. Whereas one of the disadvantages of the SLHX is the suction gas density drop due to the suction gas superheating, the SLHX contributes to a lower volumetric capacity. However, these two opposing effects resulted in a lower compressor speed with the same refrigerant mass fractions and capacity.

When the performance of R32/R152a with SLHX is compared with that of R32/R134a, the R32/R134a mixtures performed better in the cooling mode above 35% of R32 mass fraction. Essentially, both mixtures with SLHX below 30% R32 mass fraction performed similarly.

5.1.2.3 Conclusion

- For R32/R152a mixtures, the mass fraction of R32 should be higher than 40% to ensure a heating capacity comparable to R22 capacity for all operating conditions.
- R32/R152a mixtures showed a significant increase in COP (14 to 20%) for cooling conditions with a counterflow heat exchange configuration.
- R32/R152a mixtures have the lowest GWP possible of the tested mixtures, about one fourth of the R22 value, and are flammable in the whole mass fraction range.
- R32/R134a mixtures showed better or equivalent performance than R32/R152a when the mass fraction of R32 was higher than 35%. Specifically, the heating performance of R32/R134a mixtures is better than that of R32/R152a due to its larger volumetric capacity at the same R32 mass fraction.
- R32/R134a mixtures are "controllable-flammable" mixtures since they are the mixture of flammable R32 and nonflammable R134a.
- The potential benefits of SLHX are a higher performance for the given R32 mass fraction. Therefore, the R32 mass fraction can be lowered until the same volumetric capacity is obtained. This reduces the flammability of R32 rich mixtures. In fact, the COP increases for both mixtures over R22 up to 20% when the mass fraction of R32 was 40% and counterflow heat exchanger is used.
- R32/R134a mixtures with SLHX showed a higher COP increase over the case without SLHX than did R32/R152a mixtures.
- R32/R134a and R32/R152a are potential alternatives for R22 since both mixtures showed a comparable performance if the appropriate mass fraction mixture is chosen and counterflow heat exchange with the heat transfer fluid is possible.

5.1.3 ORNL Breadboard Heat Pump

The following text is based on excerpts of the earlier mentioned papers. The full texts are available from Vineyard, 1988; Vineyard and Conklin, 1991; Conklin and Vineyard, 1991; and Vineyard, Conklin, and Brown, 1993.

5.1.3.1 Test Facility

ORNL's breadboard heat pump is very similar to that of NIST's breadboard heat pump I, shown in Figure 5.1.1, except no evaporator calorimeter enclosure is used in ORNL's setup. ORNL's breadboard used water as the heat source- and sink-fluid and coaxial tube-in-tube counterflow heat exchangers for both heat exchangers.

TABLE 5.3
Test Conditions

Variable	Cooling Mode	Heating Mode
Evaporator inlet water temperature [°C]	26.7	5.6/15.6
Condenser inlet water temperature [°C]	29.4/38.9	15.6
Evaporator-side water temperature difference [°C]	20	5.6
Condenser-side water temperature difference [°C]	20	11.1

Source: Vineyard, 1988.

5.1.3.2 R13B1/R152a Test

Vineyard (1988) experimentally investigated the effects of R13B1 mass fraction of R13B1/R152a mixture on capacity control. Vineyard chose the R13B1/R152a mixture for his study because the boiling points of the pure components are moderate (not exceeding 16.7°C), recommended by Radermacher (1986) as a temperature glide for the heat pump application efficiency not to deteriorate. The R13B1 mass fraction of R13B1/R152a mixtures was varied at 25, 50 and 75% for cooling and heating test conditions as specified in Table 5.3. Both heat exchangers had the same geometry: refrigerant tube outside diameter, 25 mm; water tube outside diameter, 14 mm; and total tube length, 3 m. Results of cooling and heating tests are summarized in Figure 5.1.22. As shown in Figure 5.1.22(a), the cooling capacity of pure R152a was 28% lower than that of R22. Therefore, decreasing the mass fraction of R13B1 for the lower ambient temperature cooling could reduce the cycling loss.

Increasing the mass fraction of R13B1 could increase the heating capacity at the lower ambient temperature heating. As shown in Figure 5.1.22(b), the heating capacity of pure R152a was 33% lower than that of R22.

Moreover, the cooling and heating COPs both increased as the mass fraction of R152a increased, as shown in Table 5.4. The COP comparison tests were conducted while maintaining equivalent inlet and outlet water temperatures in the evaporator and condenser for both R22 and the mixtures. In conclusion, Vineyard experimentally confirmed that the cooling and heating capacity could be modulated by 25 to 28% by the mass fraction shift. Moreover, the estimated annual energy consumption savings was approximately 7 to 12% by decreasing the cycling losses.

5.1.3.3 R143a/R124 Test

Vineyard and Conklin (1991) experimentally investigated the effects of heat exchanger surfaces on the performance of zeotropic mixture R143a/R124 (75/25 wt.%) to improve refrigerant side heat transfer coefficients. In their study, three enhanced tubes — fluted, finned and microfin — were compared to the smooth tube. Detailed geometry of theses tubes is summarized in Figure 5.1.23 and Table 5.5. Test conditions used were based on DOE cooling conditions (27.8°C). Reference pressures were 700 kPa (12°C saturation temperature) for the evaporator and 1700 kPa

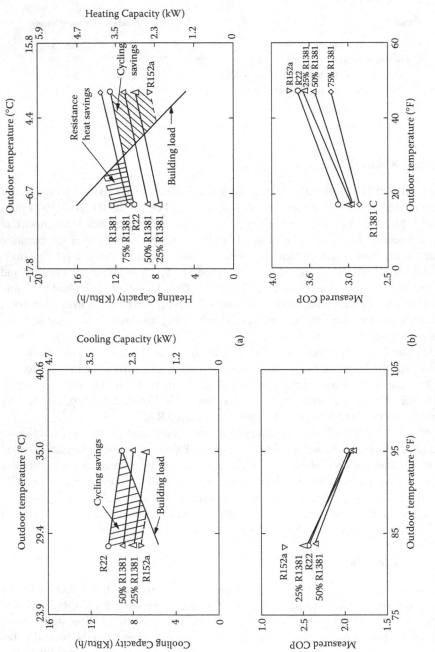

FIGURE 5.1.22 Cooling performance of R13B1/R152a (Vineyard, 1988). (a) Cooling. (b) Heating.

TABLE 5.4
Test Results

Refrigerant	R22	R152a	R13B1/R152a (25/75 wt.%)	R13B1/R152a (50/50 wt.%)	R13B1/R152a (75/25 wt.%)
Cooling COP	2.4	2.7	2.5	2.3	—
Heating COP	3.7	3.7	3.6	3.5	3.3

Source: Vineyard, 1988.

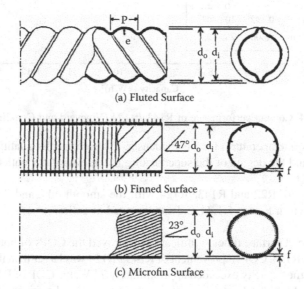

(a) Fluted Surface

(b) Finned Surface

(c) Microfin Surface

FIGURE 5.1.23 Three enhanced surface tubes (Vineyard and Conklin, 1991).

TABLE 5.5
Geometry of Enhanced Tubes

Tube Type	Smooth Tube	Fluted Tube	Finned Tube	Microfin Tube
Outside diameter [mm]	15.8	18.9	15.8	14.8
Thickness [mm]	1.0	1.1	1.1	0.5
Other [mm]	—	Spiral indentation: 0.8 Pitch: 12.7	Fin height: 0.432	Fin height: 0.305
Heat exchanger length (evaporator/condenser) [m]	2.75/1.94	2.39/1.94	2.39/1.94	2.39/1.94

Source: Vineyard and Conklin, 1991.

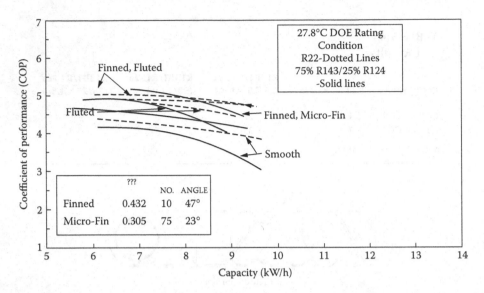

FIGURE 5.1.24 Cooling performance of R143a/R124 (Vineyard and Conklin, 1991).

(45°C saturation temperature) for the condenser. Evaporator inlet quality was approximately 0.25 and the degree of the superheating at the evaporator outlet was approximately 3 to 5°C.

Test results of R22 and R143a/R124 with the smooth tube and three enhanced tubes are shown in Figure 5.1.24 and summarized as follows:

- Enhanced surface tubes significantly improved the COPs of both R22 and R143a/R124. COP improvement of R143a/R124 was larger than that of R22.
- When the capacity exceeded approximately 7 kW, the COP of R143a/R124 dropped rapidly due to the greater pressure drop and refrigerant side heat transfer degradation associated with zeotropic mixtures.
- Different enhanced tubes worked better for the evaporator and condenser. Using a fluted tube for the evaporator and a finned tube for the condenser resulted in the best performance improvement (11 to 23%) over R22.

5.1.3.4 R143a/R124 and R32/R124 Test

As a continuing effort from Vineyard and Conklin (1991) to improve system COP, Vineyard et al. (1993) experimentally investigated three methods of improving heat transfer enhancements, inserts, finned tube and a segmented evaporator.

Inserts

Perforated foil inserts were used inside of the inner tube. The perforated brass foil ($w = 0.23$ mm, $t = 0.05$ mm, 50 holes/cm^2) was used for the second through fourth evaporator passes to enhance the nucleate boiling for the low quality region. The aluminum bent tabs, shown in Figure 5.1.25, were inserted in the second through seventh condenser passes.

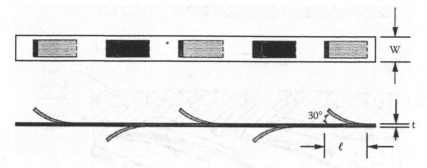

FIGURE 5.1.25 Geometry of bent-tab insert (w = 13.33 mm, t = 0.8 mm, ℓ = 2.75 m long) (Vineyard et al., 1993).

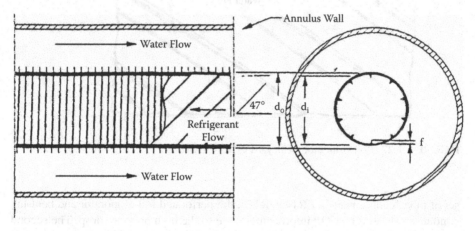

FIGURE 5.1.26 Geometry of finned tube (d_o = 15.8 mm, d_i = 13.7 mm, f = 0.38 mm) (Vineyard et al., 1993).

Finned Tube

A commercially available finned tube was used for the refrigerant side tube, as shown in Figure 5.1.26.

Segmented Evaporator

The segmented evaporator was designed to enhance the nucleate boiling heat transfer for the low vapor quality region and the convective boiling heat transfer for the high vapor quality region. Evaporator passes 1 through 4 used a rolled-fin surface as shown in Figure 5.1.27 and evaporator passes 5 through 8 used a finned tube surface as shown in Figure 5.1.26. Two sections then formed the segmented evaporator, as shown in Figure 5.1.28.

Test results of four tube combinations for the evaporator and condenser are shown in Figure 5.1.29. As shown in (a), the bent-tab inserts and the finned tube cases were compared to the smooth tube case for R22 and R143a/R125. The smooth-tube performance of R143a/R125 was 13 to 27% lower than that of R22. The first

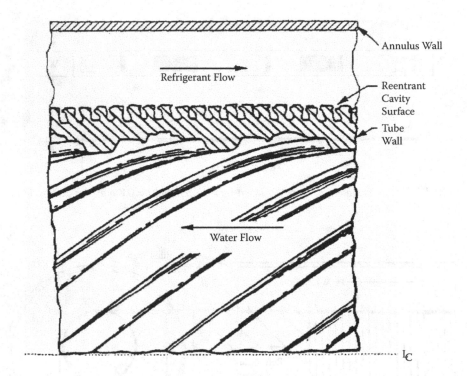

FIGURE 5.1.27 Rolled-fin tube (Vineyard et al., 1993).

set of tube enhancements for R143a/R125, the perforated foil evaporator and bent-tab condenser, showed no COP improvement due to the high pressure drop. The second set of tube enhancements for the R143a/R125, the perforated foil evaporator and finned tube condenser, showed a 15% COP improvement over the R22 at the lowest capacity and 6% lower COP than that of R22 at the highest capacity. This COP decrease for the higher capacity was due to the increased pressure drop at the higher refrigerant mass flow rate. The third set of the combination was the segmented evaporator and finned condenser. For the third set, R143a/R124 showed a 5 to 9% COP improvement over R22 and R32/R124 showed a 9 to 17% COP improvement over R22. When the third set was compared to the smooth tube case, R22 showed a 22 to 27% improvement and R143a/R124 showed a 62 to 80% improvement.

In summary, the following conclusions were obtained.

- The inserts were ineffective to improve the COP of zeotropic mixtures, R143a/R124 and R32/R124.
- Using the surface enhancement, especially the segmented evaporator and finned condenser pair, improved the COP of zeotropic mixtures relative to R22.
- R32/R124 showed almost twice better performance than R143a/R124.

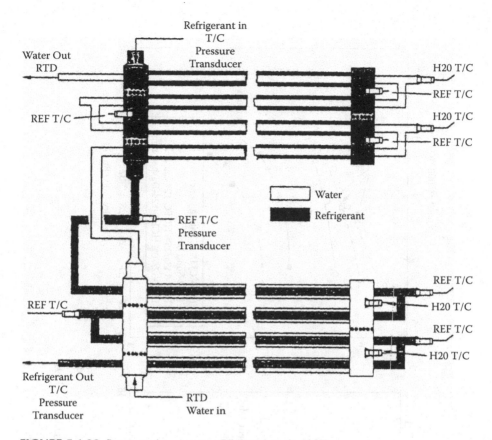

FIGURE 5.1.28 Segmented evaporator (Vineyard et al., 1993).

5.2 ACTUAL SYSTEM TESTS EXPERIENCE

This section describes experimental investigations for the performance of refrigerant mixtures in actual systems conducted at CEEE (Hwang et al., 1995) and NIST (Mulroy and Didion, 1986; Rothfleisch and Didion, 1993).

5.2.1 HEAT PUMP TEST AT CEEE

5.2.1.1 Test Facility

A test facility called a *psychrometric calorimeter* was set up to measure the steady state performance of an air-to-air heat pump. The test facility consisted of a closed indoor loop, an environmental chamber and a data acquisition system. The closed indoor loop was equipped with devices that measured the dry bulb and dew point temperatures of the air and a nozzle that measured the air flow rate downstream of the test unit, as shown in Figure 5.2.1. The test chamber was equipped with an air handling unit, as shown in Figure 5.2.2, so it could simulate the summer and winter climate conditions defined by ASHRAE Standard 116, Table 5.6 (ASHRAE, 1995).

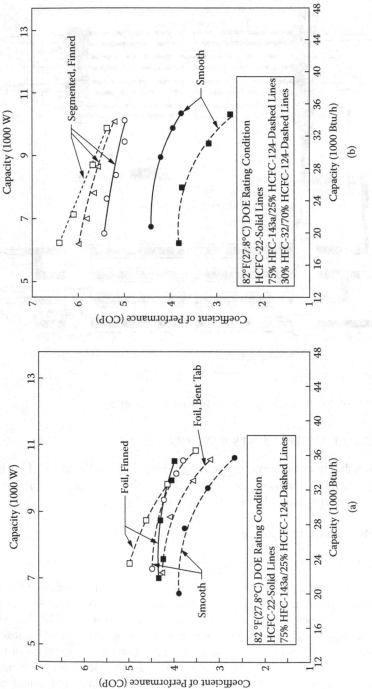

FIGURE 5.1.29 Cooling performance enhancement (Vineyard et al., 1993).

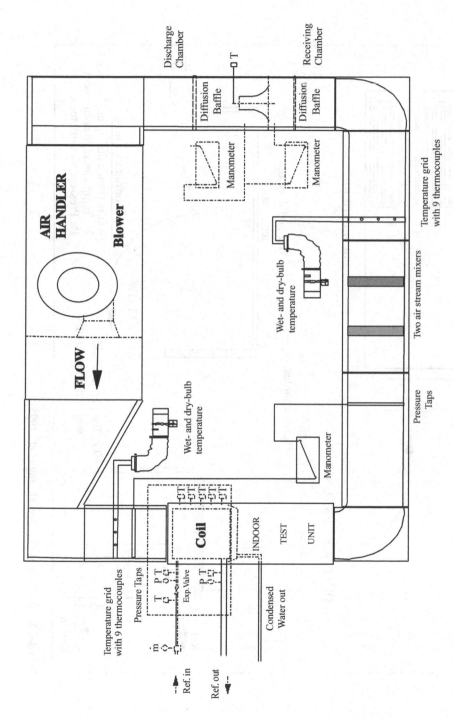

FIGURE 5.2.1 Closed indoor loop (Hwang et al., 1995).

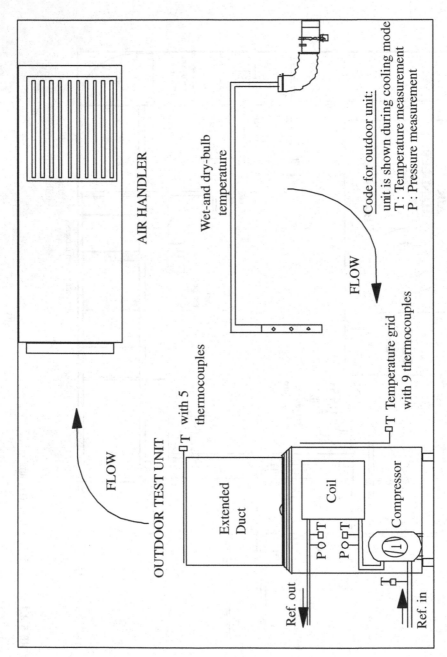

FIGURE 5.2.2 Outdoor chamber (Hwang et al., 1995).

TABLE 5.6
ASHRAE Test Conditions

Test	Indoor Dry Bulb (°C)	Indoor Wet Bulb (°C)	Outdoor Dry Bulb (°C)	Outdoor Wet Bulb (°C)
A	26.7	19.4	35.0	23.9
B	26.7	19.4	27.8	18.3
C	26.7	<13.9	27.8	18.3
D	26.7	<13.9	27.8	18.3
High temperature (47S)	21.1	15.6	8.3	6.1
Cyclic (47C)	21.1	15.6	8.3	6.1
Frost accumulation (35F)	21.1	15.6	1.7	−1.1
Low temperature (17L)	21.1	15.6	−8.3	−9.4

Source: ASHRAE, 1995.

5.2.1.2 Test Unit

The test unit was a 7.0 kW capacity split heat pump system using a reciprocating compressor and two expansion devices. One expansion device was a thermostatic expansion valve (TEV) for the heating mode and the other was a short tube restrictor (ST) for the cooling mode. Figure 5.2.3 shows the heat exchange configuration for the indoor coil. The indoor unit was installed at an inclination of 20° from the vertical with downward airflow, as shown in Figure 5.2.3. The test unit was designed for

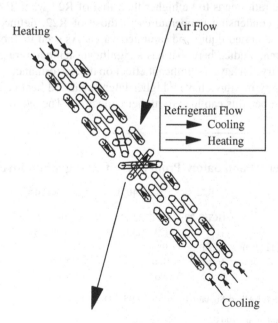

FIGURE 5.2.3 Indoor-side heat exchanger (Hwang et al., 1995).

R22; therefore, the original components were used for the evaluation of the refrigerants having a similar vapor pressure with R22. Except for the ST, the R22 tests were carried out without making any modifications to the system while the mixture tests were carried out after changing the lubricant for the compressor from mineral oil to ester oil and changing the filter drier. When testing higher vapor pressure refrigerants than R22, the compressor and TEV designed for higher operating pressures were used. As compared to the original compressor, calorimeter tests performed by the compressor manufacturer showed that the high pressure compressor had 4% lower efficiency at 54.4°C condensing temperature and 7% higher efficiency at 37.8°C condensing temperature.

5.2.1.3 Refrigerants Investigated

Several hydrofluorocarbon (HFC) refrigerants have been proposed by refrigerant manufacturers to replace R22, frequently used as a refrigerant in air-conditioning and heat pump systems. In this study, three replacements were investigated: a ternary mixture of R32/R125/R134a in the mass fraction of 23/25/52 wt.% (R407C) and two binary mixtures, R32/R125 in the mass fraction of 50/50 wt.% (R410A) and R32/R125 in the mass fraction of 45/55 wt.% (R410B). These refrigerants are non-ozone-depleting substances, as shown in Table 5.7. With the exception of the temperature glide, R407C has similar thermodynamic properties to those of R22, as shown in Table 5.8. Therefore, similar size hardware, such as the compressor and the heat exchangers, can be utilized. The vapor pressure curve of R407C is close to that of R22, as shown in Figure 5.2.4, but it is slightly higher at higher temperatures and slightly lower at lower temperatures than that of R22.

The pressure ratio tends to be higher than that of R22 when R407C has similar evaporating and condensing temperatures to those of R22, defined as the average temperatures of saturated liquid and saturated vapor. As seen in Table 5.5, R407C is the only refrigerant studied here that has a significant temperature glide, so the heat exchange geometry can have a significant effect on the performance of this refrigerant.

The two binary mixtures have a higher latent heat and heat capacity than those of R22 and these benefits can lead to higher capacities. The saturated vapor density

TABLE 5.7
Environmental and Safety Properties of Refrigerants Investigated

Refrigerant	R22	R407C	R410A	R410B
Components	HCFC-22	HFC-32/125/134a (23/25/52 wt.%)	HFC-32/125 (50/50 wt.%)	HFC-32/125 (45/55 wt.%)
O.D.P. (CFC-11=1.0)	0.055[a]	0[a]	0[b]	0[c]
G.W.P. (CO$_2$ = 1.0)	1600[a]	1600[a]	2200[b]	2020[c]
Flammability	No	No	No	No

[a] DuPont, 1994; [b] Allied Signal Chemicals, 1995; [c] DuPont, 1995.

Source: Hwang et al., 1995.

TABLE 5.8
Thermodynamic Properties of Refrigerants Investigated
(Hwang et al., 1995)

Refrigerant		R22	R407C	R410A	R410B
Molecular weight	[g/mol]	86.5	86.21[a]	72.62[b]	75.63[c]
Normal boiling point	[°C]	−40.9	−43.61[a]	−52.72[b]	−51.83[c]
Critical temperature	[°C]	96.2	86.71[a]	72.52[b]	71.63[c]
Critical pressure	[bar]	50.5	46.21[a]	49.52[b]	47.83[c]
Latent heat	at 25°C [kJ/kg]	180.6	189.3	194.0	184.9
Bubble pressure	at 25°C [bar]	10.4	11.9	16.5	16.4
Liquid density	at 25°C [kg/m³]	1192	1154	1083	1096
Saturated vapor density	at 25°C [kg/m³]	44.5	42.5	62.2	64.5
Temperature glide	at 1 atm [K]	0.0	7.1	0.1	0.1
Sat. liquid heat capacity	at 10°C [kJ/kg K]	1.23	1.41	1.48	1.46
Sat. liquid heat capacity	at 50°C [kJ/kg K]	1.47	1.68	1.92	1.89
Sat. vapor heat capacity	at 10°C [kJ/kg K]	0.81	0.88	0.94	0.94
Sat. vapor heat capacity	at 50°C [kJ/kg K]	1.17	1.20	1.54	1.55

[a] DuPont, 1994; [b] Allied Signal Chemicals, 1995; [c] DuPont, 1995.

Source: Gallagher et al., 1993, unless otherwise noted. Hwang et al., 1995.

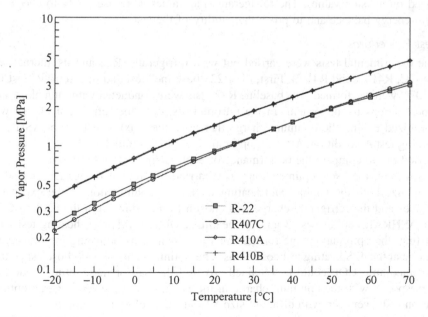

FIGURE 5.2.4 Vapor pressures.

of these refrigerants is approximately 44% higher than that of R22, which also translates into higher capacities for the same size compressor. To test the fluids with the original heat exchangers, the displacement of the compressor had to be reduced to a similar capacity. These binary mixtures have a higher vapor pressure by approximately 60% compared to R22 but the slope of the vapor pressure curve is quite similar to that of R22 in the temperature range of interest (Figure 5.2.4). Therefore, the pressure ratio of these refrigerants is expected to be similar to that of R22.

5.2.1.4 Test Procedure

Capacity Measurement

The experiments to measure the capacity and COP were performed based on ASHRAE Standard 116 (ASHRAE, 1995) and ARI Standard 210/240 (ARI, 1994). In this study, the air side capacity and the refrigerant side capacity were measured. The loop air enthalpy method was used to measure the air side capacity: the air flow rate and air enthalpy difference between inlet and outlet of the indoor coil were measured. A nozzle apparatus measured the air flow rate. In measuring the refrigerant side capacity, the refrigerant mass flow rate and the refrigerant enthalpies at the inlet and outlet of the heat exchanger were used. The mass flow rate was measured by a Coriolis-type mass flowmeter and the refrigerant enthalpies were calculated by REFPROP Version 4.0 (Gallagher et al., 1993) from the temperature and pressure measurements.

ASHRAE Standard 116 (1995) requires that the capacities determined using these two methods agree within 6% of each other. The two methods agreed within 3% for all tests conducted in this study. The reported capacity and COP values were based on air side values. The refrigerant side values were used only to check the total energy balance and to prove the validity of the test.

Test Procedures

The experimental tests were carried out with refrigerants R22 and its alternatives, R407C, R410A and R410B. First, the R22 "baseline" test and the "retrofit" test for R407C were performed. The baseline R22 tests were conducted without making any modifications to the system. In the retrofit tests, only the refrigerant charge was optimized to find the optimum charge that gave the maximum COP at ASHRAE cooling test A condition. After the optimum charge was found, ASHRAE tests were carried out to compare the performance of each refrigerant and mixture.

Second, the "soft optimization" was carried out for all refrigerants tested to maximize both the cooling and heating COP. The combination of the expansion devices and the refrigerant charge was chosen to maximize both the cooling COP for ASHRAE cooling test B and the heating COP for ASHRAE heating test 47S. At first, the optimum charge for heating was obtained by adjusting the refrigerant charge at the 47S heating test conditions. The optimum charge was chosen such that the maximum COP could be obtained. To find the ST for the cooling mode that corresponds with the optimum charge in the heating mode, several charge optimization tests were run with different size STs at the cooling test conditions.

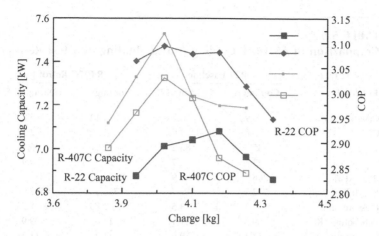

FIGURE 5.2.5 Charge optimization of R22 and R407C (Hwang et al., 1995).

5.2.1.5 Test Results

R22 Baseline Test

The charge optimization test results for R22 are shown in Figure 5.2.5. When increasing the charge, the capacity and COP for R22 increase to a maximum point and then decrease. COP for R22 reaches a maximum at approximately 4.0 kg charge and the capacity reaches a maximum at 4.2 kg charge. By increasing the charge, the amount of subcooling increases while the superheat decreases. Throughout the ASHRAE tests, the refrigerant charge was maintained at 4.0 kg, corresponding to the maximum COP for ASHRAE Cooling Test A.

R407C Retrofit Test

The mixture R407C was first tested as a "retrofit." Only the refrigerant charge was optimized to obtain maximum COP. The same procedure used for R22 was used for the charge optimization test of R407C and the results are shown in Figure 5.2.5. The COP is at maximum at approximately 4.0 kg, the same as that of R22. After these tests, ASHRAE tests were carried out with the optimum charge. The results for both R22 and R407C are listed in Tables 5.9 and 5.10.

Comparison of R22 Baseline, R407C Retrofit

The retrofit test results for R407C are compared with R22 baseline test results in Table 5.9 for cooling test A and heating test 47S. In Table 5.10, the test results for ASHRAE cooling test B and heating test 17L are compared. The retrofit test results for R407C show a 2.6% higher capacity and a 4.5% lower COP in the cooling test A case and a 6.4% lower capacity and a 16.5% lower COP in the heating test 47S case as compared to the R22 baseline (Table 5.9). Cooling test B has similar results to cooling test A case (Table 5.10) while heating test 17L has a greater degradation than the heating test 47S case (Table 5.10). Although R407C has similar thermodynamic characteristics as R22, some key cycle parameters are different. R407C shows a 16.5% higher pressure ratio compared to that of R22 for the cooling A case.

TABLE 5.9
Comparison of ASHRAE Cooling A and Heating 47S Test Results

Test Type	R22 Baseline		R407C Retrofit	
Test Mode	Cooling A	Heating 47S	Cooling A	Heating 47S
Capacity [kW]	6.95	6.24	7.13	5.84
COP	3.08	3.03	2.94	2.53
$T_{discharge}$ [°C]	85.5	83.2	82.0	96.5
$T_{suction}$ [°C]	22.6	8.7	21.8	9.9
P_{cond} [kPa]	1768.7	1793.9	2124.2	2678.9
P_{evap} [kPa]	712.7	507.4	736.2	502.9
Pressure ratio	2.48	3.54	2.89	5.33
Subcooling [K]	9.6	21.1	14.0	39.9
Superheating [K]	11.1	9.9	11.3	11.3

Source: Hwang et al., 1995.

TABLE 5.10
Comparison of ASHRAE Cooling B and Heating 17L Test Results

Test Type	R22 Baseline		R407C Retrofit	
Test Mode	Cooling B	Heating 17L	Cooling B	Heating 17L
Capacity [kW]	7.12	3.40	7.13	2.69
COP	3.52	2.09	3.22	1.59

Source: Hwang et al., 1995.

Also, the cycle shifts to higher evaporating and condensing pressures. The higher pressure ratio lowers the compressor efficiency. Although R407C has a slightly higher capacity; it has a lower COP.

R407C retrofit shows more degradation for heating than for cooling. This can be explained by the heat exchange configuration of the indoor coil as shown in Figure 5.2.3. The heat exchanger configuration contributes to the capacity degradation in the heating mode, especially for mixtures that have a temperature glide.

Soft Optimization Test Results

To improve the steady-state performance of the mixture, the soft optimization was carried out for R22, R407C, R410A and R410B that could reconcile the imbalance between the cooling and heating optima. As already mentioned, the optimum charge was found for the heating test 47S, then the ST was adjusted so that the optimum-cooling COP occurs at the same charge. When the optimum charge was the same for different sizes of STs as for R410B, the subcooling and superheating were evaluated to choose the optimum ST size. For R410B, a 1.55 mm ST has the wider range of refrigerant charge with acceptable subcooling and superheating than those of 1.60 mm ST. Therefore, the 1.55 mm ST was chosen for R410B and the result

TABLE 5.11
Optimum Charge for Each Operating Mode

Refrigerant	R22	R407C	R410A	R410B
Heating optimum charge [kg]	3.6	3.4	3.6	3.4
Cooling optimum charge [kg]	3.6	3.4	3.6	3.8
ST size [mm]	1.65	1.70	1.55	1.55

Source: Hwang et al., 1995.

of this procedure is shown in Table 5.11. Except for R410B, it was possible to adjust the charge and ST so the optimum charges for heating and cooling test conditions were the same. For R410B, the average between the optimum heating and cooling charges was used. The larger ST, 1.65 mm, was also tested but the reasonable values of subcooling and superheating were not obtained. This mismatch could have been eliminated or minimized if the more finite increases for the size of STs and refrigerant charges were tested.

Steady State Performance Results

After the soft optimization tests, ASHRAE tests were carried out with the optimum refrigerant charge and ST. Test results for capacity and COP are compared with other data in Tables 5.12, 5.13 and 5.14. R407C had approximately the same cooling and heating capacity (within 3% of R22) but also a slight degradation in cooling and heating COP (4 to 8%). The existing heat exchanger was not suited for utilizing the temperature glide of this mixture and contributed to the slight degradation in capacity. As expected, this refrigerant had a larger pressure ratio and this also might contribute to the degradation in COP.

The two binary mixtures showed higher cooling capacity by up to 7% and COP up to 4% compared to those of R22 despite the smaller compressor. These refrigerants also had better heating capacity by 3 to 6% at 47S and up to 20% in the 17L

TABLE 5.12
ASHRAE Cooling A Test Results

Refrigerant	R22	R407C	R410A	R410B
Capacity [kW]	6.82	6.89	7.16	7.07
COP	3.20	3.06	3.25	3.23
T_{dis} [°C]	76.3	78.1	75.7	75.0
T_{suc} [°C]	18.0	23.3	20.5	20.8
P_{cond} [kPa]	1641.8	1825.9	2611.8	2587.3
P_{evap} [kPa]	707.5	704.6	1097.1	1081.1
Pressure ratio	2.32	2.59	2.38	2.39
Subcooling [K]	4.0	4.8	2.8	3.7
Superheating [K]	6.7	14.3	10.0	10.6

Source: Hwang et al., 1995.

TABLE 5.13
ASHRAE Heating 47S Test Results

Refrigerant	R22	R407C	R410A	R410B
Capacity [kW]	6.13	5.95	6.51	6.29
COP	3.13	2.96	3.02	3.03
T_{dis} [°C]	81.4	77.1	71.6	65.5
T_{suc} [°C]	9.2	8.9	2.0	2.1
P_{cond} [kPa]	1678.5	1860.3	2618.0	2435.1
P_{evap} [kPa]	494.6	478.9	826.3	812.8
Pressure ratio	3.39	3.88	3.17	3.00
Subcooling [K]	20.2	21.4	17.1	10.5
Superheating [K]	9.5	11.7	0.8	1.2

Source: Hwang et al., 1995.

TABLE 5.14
ASHRAE Cooling B and Heating 17L Test Results

Refrigerant	R22	R407C	R410A	R410B
B capacity [kW]	7.34	7.16	7.83	7.64
B COP	3.78	3.49	3.92	3.88
17L capacity [kW]	2.77	2.75	3.32	3.33
17L COP	1.97	1.82	1.87	1.89

Source: Hwang et al., 1995.

test. R410A and R410B showed similar cooling and heating performances when compared with each other.

As expected, these refrigerants have higher evaporating and condensing pressures by approximately 60% compared with those of R22 and the pressure ratio was nearly the same as that for R22.

The results of this study for R407C and R410B are compared with the results presented by AREP (Alternative Refrigerants Evaluation Program) (Godwin, 1993) in Table 5.15, while the results of R410B are compared with the results by Kolliopoulos (1994). Overall, the data of this study agree well with the AREP data. For R410B, the data show good agreement with those of Kolliopoulos except for the low temperature heating capacity, which is higher than that of Kolliopoulos. This significant gain in capacity arises because at higher ambient temperatures and therefore higher capacities, the heat exchangers are slightly undersized for the high pressure fluids. This occurs because it was not possible to exactly match the high pressure compressor's capacity to that of the existing system. Support for this conclusion comes from the minimal superheating.

For the data presented here, the TEV maintained the superheating as low as 1 to 2°C while in the Kolliopoulos data, the superheating was maintained at 11.1°C.

TABLE 5.15
Comparison of Results with AREP Data

Refrigerant	Source	Cooling Capacity	Cooling COP	Heating Capacity	Heating COP
R407C	Present study	0.98–1.01	0.92–0.96	0.97–0.99	0.92–0.95
	AREP	0.93–1.01	0.90–0.97	0.98–1.02	0.93–1.02
R410A	Present study	1.05–1.07	1.02–1.04	1.06–1.20	0.95–0.97
	AREP	0.98–1.05	1.01–1.06		
R410B	Present study	1.04	1.01–1.03	1.03–1.20	0.96–0.97
	Kolliopoulos, 1994	0.99–1.00	0.98–1.00	0.99–1.03	1.03–1.09

Note: The above data are the ratios of each refrigerant performance to R22 performance.

Source: Hwang et al., 1995.

Another difference is that a reciprocating compressor was used in this study and a scroll compressor was used in the Kolliopoulos study.

Seasonal Performance Results

When evaluating the overall performance of heat pumps, including the steady state and cyclic performances, the seasonal performance factors seasonal energy efficiency ratio (SEER) and heating seasonal performance factor (HSPF) are used. These factors are important because they approximate operation over an entire season. The seasonal performance was calculated based on ASHRAE Standard 116 (ASHRAE, 1995). Eight different ASHRAE test results, including the steady state and cyclic performances, were used in this calculation.

The seasonal performance factors for each refrigerant are calculated and compared in Table 5.16. Relative to R22, R407C has a 6.8% lower SEER and a 4.7% lower HSPF, while R410A has a 2.6% higher SEER and nearly the same HSPF and R410B has a 1.7% higher SEER and a 1.8% higher HSPF. Two binary mixtures have lower COP for 47S test conditions but their better cyclic performance contributes to having a slightly better HSPF.

R407C, having similar operating pressures as R22, has the advantage of being a *retrofit*, which means minimal changes to the system for a quick switch to alternatives. Only the expansion devices and the refrigerant charge must be chosen carefully to have performance within 7.8% of R22. R407C has a 6.8% lower cooling seasonal performance and a 4.7% lower heating seasonal performance as compared to R22. The reduction in performance as compared to R22 is relatively small and can be overcome with systems specifically designed for R407C.

The two binary refrigerants require careful design considerations, especially heat exchangers, pipes and compressors, due to the much higher operating pressures. Although these binary refrigerants need significant system modification, they have the benefit of higher steady state and seasonal performances. These benefits could be even greater if the system is designed specifically for these fluids. Overall, the binary fluids perform better than R22. R410A has a cooling seasonal performance

TABLE 5.16
Seasonal Performance Test Results

Refrigerant	R22	R407C	R410A	R410B
SEER [Btu/kW-hr]	11.7	10.9	12.0	11.9
HSPF	1.70	1.62	1.70	1.73

Notes:
1. Conditions in SEER calculation.
   (a) Oversize factor: 10%.
   (b) Design temperature of 35°C (needs eight temperature bins ranging from 18 to 41°C).
   (c) The bin hours are based on the U.S. national average climate.
   (d) The building load (BL) was selected from the R22 baseline case.
2. Conditions in HSPF calculation.
   (a) The regional outdoor temperature: –22°C.
   (b) Location requires 18 temperature bins ranging from –32 to 18°C.
   (c) The bin hours were based on Region V in ARI Standard 210/240 (1994).

Source: Hwang et al., 1995.

that is 2.6% better than R22 and a heating seasonal performance that is essentially the same as R22. R410B has cooling and heating seasonal performances that are 1.7 and 1.8% better than R22, respectively.

5.2.1.6 Effects of Heat Exchanger Geometry on Mixture Performance

As explained in Section 3.8, the temperature glide of zeotropic mixtures can be utilized in counter flow heat exchange. Counter flow heat exchange can reduce the average temperature difference between the two heat exchanging streams so the cycle can approach a Lorenz cycle, resulting in reduction of the thermodynamic external irreversibility and consequently having a higher heat exchange efficiency. It is very important to achieve the counter flow heat exchange configuration for the heat exchanger when using zeotropic mixtures.

Path Modification Test (R407C only)

To implement the benefit of a temperature glide of the zeotropic mixture refrigerant R407C, the path modification option was examined. The path modification requires altering the refrigerant path to attain a near counter flow configuration in the indoor coil for the heating mode. The indoor coil of the test unit has a quasi-cross flow heat exchange configuration. In the cooling mode, the refrigerant flows from bottom to top, whereas in the heating mode the refrigerant flows from top to bottom, as shown in Figure 5.2.3. The heat exchange pattern between air and refrigerant is near counter flow for the cooling mode but it is near parallel flow for heating. Installing four check valves, as shown in Figure 5.2.6, changed the indoor coil path to supply the refrigerant to the indoor coil in the same direction regardless of the action of the four-way valve. With this path modification, the refrigerant could be supplied into the indoor coil from bottom to top for both cooling and heating modes to maintain a near counter flow heat exchange between air and refrigerant at all times. This option is important when using zeotropic mixtures that have a temperature glide because the refrigerant path is reversed depending on the operating mode in a heat pump.

Path modification results show a 3.7% lower capacity and a 5.5% lower COP for cooling test A as compared to that of R22, as shown in Table 5.17. The four additional check valves cause an additional pressure drop, resulting in a lower mass flow rate and a lower evaporating pressure. But it shows a significant improvement in the heating COP (10.3%) for heating test 47S, as compared to the retrofit case shown in Table 5.18. The greatest improvement for path modification can be observed in the low temperature heating test 17L: 15.6% for capacity and 17.6% for COP, as shown in Table 5.19.

The refrigerant temperatures along the cycle for retrofit and path modification are compared in Figures 5.2.7 and 5.2.8 for cooling test A and heating test 47S. In these figures, the x-axis is not to scale; it indicates just the sequence of each temperature probe as the refrigerant passes on its way through the cycle. The first point is the compressor discharge and the last point is the compressor inlet. The profile for the cooling of R407C is similar to that of R22. The heating case shows a higher compressor discharge temperature and a lower evaporating temperature than that of R22, resulting in a high pressure ratio for R407C. The temperature profiles in the evaporator and condenser show large differences. These temperature profiles explain the performance degradation of the retrofit case, which had the large temperature difference because of the near parallel flow configuration. They also explain the performance improvement due to path modification, which resulted in a small temperature difference because of the near counter flow configuration. The path modification can improve the performance by reducing the thermodynamic irreversibility. Although path modification improved the performance of the R407C over the retrofit, it still experiences degradation as compared to R22, especially in the heating mode: a 4.2% lower capacity and a 7.9% lower COP at heating test 47S. The reason for this degradation is the retrofit was performed after the charge was optimized at cooling test A conditions without optimization at the heating test conditions, so degradation of the heating performance is natural. These results indicate that more extensive optimization is necessary, such as the soft optimization.

5.2.2 Heat Pump Test at NIST

Manufacturers must take care of existing products in the field and products to be shipped until new systems replace them. Therefore, an implementation of short-term alternatives is as important as the long-term alternatives. "Retrofit" (or drop-in) refrigerants are short-term alternatives since they only require minor system changes.

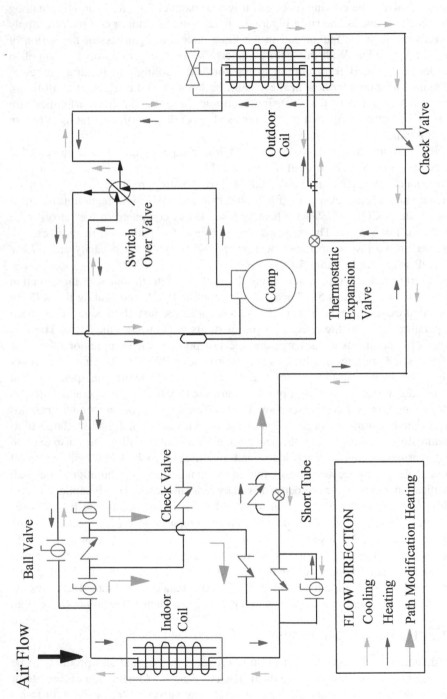

FIGURE 5.2.6 Schematic cycle diagram of path modification (Hwang et al., 1995).

TABLE 5.17
Comparison of ASHRAE Cooling A Test Results

Case	R22 Base	R407C Retrofit	R407C Path Modification
Capacity [kW]	6.95	7.13	6.69
COP	3.08	2.94	2.91
T_{dis} [°C]	85.5	82.0	77.8
T_{suc} [°C]	22.6	21.8	19.6
P_{cond} [kPa]	1768.7	2124.2	1959.6
P_{evap} [kPa]	712.7	736.2	699.8
Pressure ratio	2.48	2.89	2.80
Subcooling [°C]	9.6	14.0	10.4
Superheating [°C]	11.1	11.3	10.8

Source: Hwang et al., 1995.

TABLE 5.18
Comparison of ASHRAE Heating 47S Test Results

Case	R22 Base	R407C Retrofit	R407C Path Modification
Capacity [kW]	6.24	5.84	5.98
COP	3.03	2.53	2.79
T_{dis} [°C]	83.2	96.5	86.3
T_{suc} [°C]	8.7	9.9	9.0
P_{cond} [kPa]	1793.9	2678.9	2236.6
P_{evap} [kPa]	507.4	502.9	483.1
Pressure ratio	3.54	5.33	4.63
Subcooling [°C]	21.1	39.9	29.9
Superheating [°C]	9.9	11.3	11.6

Source: Hwang et al., 1995.

TABLE 5.19
Comparison of ASHRAE Cooling B and Heating 17L Test Results

Case	R22 Base	R407C Retrofit	R407C Path Modification
"B" capacity [kW]	7.12	7.13	7.15
"B" COP	3.52	3.22	3.40
"17L" capacity [kW]	3.40	2.69	3.11
"17L" COP	2.09	1.59	1.87

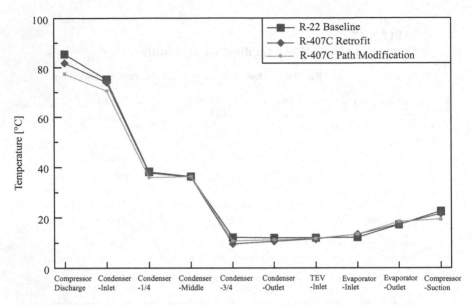

FIGURE 5.2.7 Refrigerant temperature along the cycle in the cooling mode (Hwang et al., 1995).

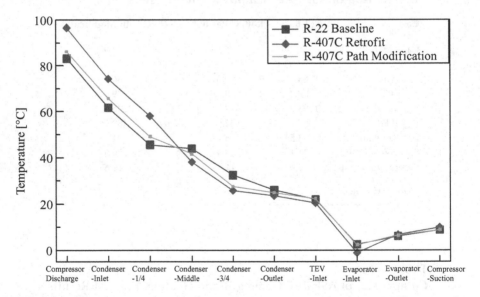

FIGURE 5.2.8 Refrigerant temperature along the cycle in the heating mode (Hwang et al., 1995).

In this section, two sets of drop-in tests conducted by NIST are introduced. The first set of tests was conducted with a zeotropic refrigerant, a R13B1/R152a (65/35 wt.%) mixture in a relatively unmodified residential heat pump designed for R22 (Mulroy and Didion, 1986). The second set of tests was conducted with two R32 mixtures,

R32/R134a and R32/R125/R134a (Rothfleisch and Didion, 1993). The following text is based on excerpts of the above-mentioned reports. The full text is available as NBSIR 86-3422 and NISTIR 93-5321.

5.2.2.1   R13B1/R152a Test

The purposes of the first set of tests were to investigate the potential benefits of zeotropic mixtures in terms of thermodynamic efficiency by utilizing gliding temperatures during phase change and capacity modulation by using mass fraction shift. To fulfill these purposes, the following four tests were conducted with a nominal 3 refrigeration ton, split system, air-to-air unitary heat pump:

1. Pure retrofit tests without any modification of the test heat pump unit.
2. Simple cycle optimization tests in terms of a refrigerant charge and orifice tube choice.
3. Effect of the zeotropic mixture mass fraction shift on the performance.
4. System with a receiver to evaluate its effect on the zeotropic mixture mass fraction shift.

Details of the experimental results are as follows.

After the first series of tests (Test 1), it was found that the retrofit tests of mixtures without any modifications for the system designed for R22 were not a valuable evaluation. Considering different thermophysical properties of each refrigerant, there was a need for system design modifications. However, minimum changes were attempted without a major system change to pursue the retrofit option. Thus, charge optimization and expansion device size are optional (Test 2). When optimizing the expansion device for cooling, the degree of suction superheating should be considered for both tests condition (Test A and Test B). In general, the degree of superheating decreases as the ambient temperature increases. Therefore, if the degree of superheating at Test B is insufficient, the compressor suction can be flooded at Test A conditions. The optimization of the expansion device for heating requires maintaining a proper degree of subcooling instead. In general, the degree of subcooling decreases as the ambient temperature decreases. If the condenser outlet becomes two-phase, a large amount of the refrigerant is being stored in the accumulator. This causes a shift in the concentration of the refrigerant to contain more low boiling refrigerant and boosts the heating capacity. Therefore, selecting an expansion device such as subcooling at Test 47 enables the system to balance the COP at high ambient temperature heating with increased heating capacity at low ambient temperatures.

As a simple cycle optimization for each refrigerant, two cooling tests, Test A and Test B, and two heating tests, Test 47S (high temperature heating) and Test 17L (low temperature heating), were performed. These tests represent the summer and winter conditions defined by ASHRAE Standard 116, Table 5.6 (ASHRAE, 1995). Test results are shown in Figure 5.2.9.

The optimum size of the orifice that gives a maximum capacity is different for Test A and Test B because the cycle parameters are different depending on the ambient temperature. For example, the optimum orifice area at Test A is 30% smaller

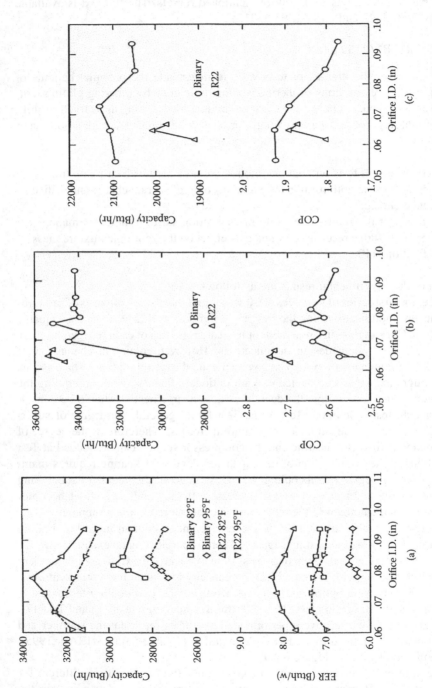

FIGURE 5.2.9 Charge and expansion device optimization test results (Mulroy and Didion, 1986). (a) R22. (b) R32/R134a (34/66 wt.%). (a) Test A and Test B. (b) Test 47S. (c) Test 17L.

TABLE 5.20
Comparison of Operating Pressures at Test A

Refrigerant	R22		R13B1/R152a	
Expansion device	Capillary tube	Orifice 2.0 mm	Capillary tube	Orifice 2.1 mm
Discharge pressure [kPa]	1917	1924	2816	2048
Discharge temperature [°C]	111	117	105	98
Suction pressure [kPa]	572	628	571	627
Pressure ratio	3.35	3.06	3.83	3.27

Source: Mulroy and Didion, 1986.

and 6% greater than at Test B for R22 and R13B1/R152a, respectively. However, those conditions providing the maximum COP at Test B determine the optimum orifice size and charge since Test B has more bin hours. Another purpose of the optimization of an expansion device is to obtain a minimal high side pressure and subcooling. Table 5.20 compares the operating pressures of two refrigerants. As can be seen from Table 5.20, the cycle with an optimized orifice has a pressure ratio reduced by 9 and 15% as compared to the drop-in tests with a capillary tube for R22 and R13B1/R152a, respectively. The suction pressure increases for the orifice cases due to the reduced degree of superheating at the evaporator outlet as an outcome of an orifice size optimization. The discharge temperature of R13B1/R152a is about 10 K lower than that of R22, as shown in Table 5.20.

As can be seen from Figure 5.2.9, the performance of the R13B1/R152a mixture was lower than that of R22 in general: a 3% loss in COP for Test 47S, a 12% loss in capacity and 11% loss in COP for the cooling test, except a 6% improvement in capacity and no change in COP for Test 17L (−8.3°C). One of the reasons for this performance degradation is an inability to utilize a temperature glide of the zeotropic mixtures. There are two barriers to utilize the temperature glide of zeotropic mixtures in an air-to-air heat pump or air conditioner: the crossflow heat exchange configuration for the most widely accepted heat exchangers and the refrigerant flow reversal when switching the operating mode from cooling to heating for heat pumps. The second reason is that these tests were conducted on a drop-in basis without any major system modification.

To increase the heating capacity at low ambient temperatures, it was necessary to sacrifice the subcooling at the condenser outlet. To investigate the performance potential of the R13B1/R152a mixture with subcooling and a large mass fraction shift, a high mass fraction test was conducted by charging a premixed mixture at 86% of R13B1. This investigation of the mass fraction shift on the capacity modulation showed that the heating capacity at −8.3°C increased 18% above that with R22 when the R13B1 mass fraction was 86% (Table 5.21). This mass fraction shift strategy would contribute to increased seasonal performance in cold climates by reducing the need for electric heat.

To realize this benefit, a receiver was added in the liquid line to create a mass fraction shift at low air temperatures. The volume of the receiver was 2.3 l, which

TABLE 5.21
Performance of High R13B1 Mass Fraction

Refrigerant	R22	R13B1/R152a (86/14 wt.%)	
Ambient temperature [°C]	–8.3	–8.3	47
Capacity [kW]	6.3	7.4	11.5
COP	1.98	2.00	2.41
Discharge pressure [kPa]	1241	1786	2124
Discharge temperature [°C]	100	81	78
Suction pressure [kPa]	296	425	665

Source: Mulroy and Didion, 1986.

could store 2.6 kg of R22 or R13B1/R152a (65/35 wt.%) mixture if completely filled. This extra amount of refrigerant in the receiver during steady-state operations resulted in a total system charge that would overfill the accumulator during cyclic operation. This large mass fraction shift to the higher mass fraction of R13B1 resulted in a large volumetric capacity by increasing the suction vapor density. Indeed this option enhances the control over the mass fraction shift by adding one more degree of freedom in terms of refrigerant storage, resulting in a 14 and 2% capacity and COP increase, respectively, at the –8.3°C rating point.

Mixtures should be evaluated on a reliability basis as well as energy performance. For the vapor compression cycle, this usually means consideration of the discharge pressure and temperature levels as well as the pressure difference and ratio. For a heat pump, the high-stress condition resulting from high pressures and temperatures will occur during the highest condenser operating conditions, which corresponds with the cooling mode.

Conclusions
- Retrofit of R22 by R13B1/R152a mixtures improved low temperature heating capacity up to 6%, but lost in high temperature heating COP by 3%, in cooling capacity by 12% and in cooling COP by 11%.
- Mass fraction shift effectively controls the heating capacity and showed 18% higher low temperature heating capacity than that of R22.
- A liquid line receiver enhanced the control over the mass fraction shift and resulted in 14 and 2% low temperature heating capacity and COP increases, respectively.
- Both energy performance and reliability should be emphasized for mixtures.

5.2.2.2 R32/R134a Test

The second set of tests was conducted with R32/R134a (34/66 wt.%). The purpose of this study was to experimentally evaluate a drop-in performance of R32/R134a and a performance of the cycle with SLHX. The mass fraction of R32/R134a was chosen such that the vapor pressure of this mixture is similar to that of R22.

Moreover, the performance of this binary mixture has shown an equivalent performance to that of R22 with a counter flow heat exchange configuration (Rothfleisch and Didion, 1993).

Retrofit Test of R32/R134a

Similar to the tests for R13B1/R152a, a retrofit test was conducted with an optimization of the expansion device opening and refrigerant charge at ASHRAE Test A condition. The optimum combination of expansion device opening and refrigerant charge were determined when that combination yields the maximum COP. The system was then tested at ASHRAE Test B condition with the optimum combination determined from Test A.

In general, the residential heat pumps have two expansion devices, one for cooling and the other for heating. The second optimization for the heating mode was conducted at the ASHRAE Test 47S condition similar to the cooling mode optimization.

The results of the cooling mode optimization are shown for R22 and R32/R134a in Figure 5.2.10. In these tests, the degree of subcooling was adjusted by changing the refrigerant charge and the degree of superheating at the evaporator outlet was adjusted by the expansion valve opening between 1.7 and 2.8°C. The highest capacity was obtained at a minimum of superheating. Therefore, for each value of subcooling, the expansion valve opening was optimized. The optimum values of subcooling were approximately 5.6 and 2.8°C for R22 and R32/R134a, respectively, as shown in Figure 5.2.10.

The same optimization procedures applied for the heating mode. The results of the heating mode optimization for R22 showed that there was no definite maximum at various values of the subcooling, as shown in Figure 5.2.11. As can be seen from this result, the maximum COP is achievable when the subcooling becomes zero. However, the degree of subcooling decreases as the ambient temperature drops. Therefore, if the subcooling were set to zero at Test 47S conditions, the condenser outlet would be two-phase at Test 17L conditions. When the ambient temperature drops, the heating load increases but the heating capacity decreases. If the condenser outlet became two-phase, the heating capacity would further decrease. Therefore, the heating optimization should be conducted while balancing the heating COP at the high ambient temperature and the heating capacity at the low ambient temperature.

Figures 5.2.12 through 5.2.14 show the effect of subcooling on the capacity, COP and condensing pressure. As the subcooling increases, the capacity increases since the latent heat of evaporation increases due to the lower evaporator inlet enthalpy. As the subcooling increases, the power increases since the condensing pressure increases as shown in Figure 5.2.14. The reason the condensing pressure increases is that as the subcooling increases, a smaller portion of the condenser is used for the two-phase and superheated gas heat transfer.

The only way to maintain the same capacity or even increase capacity is to increase the temperature difference if the condenser UA is constant. In other words, the condensing temperature and condensing pressure increase and result in an increased pressure ratio. When the compression pressure ratio increases, the compressor delivers less mass flow rate and the power per unit refrigerant mass increases.

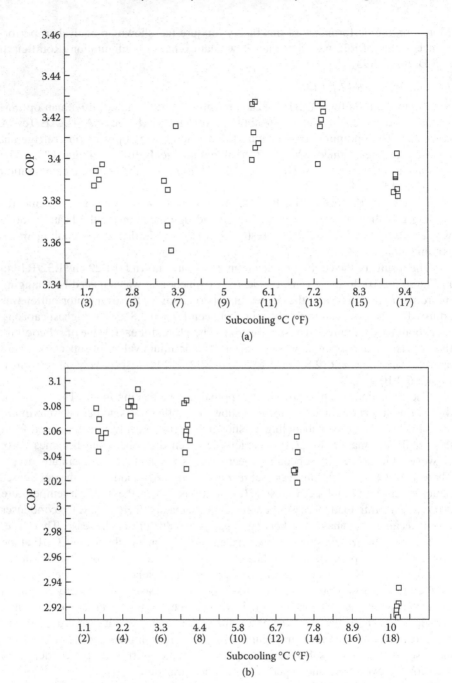

FIGURE 5.2.10 Cooling mode optimization (Rothfleisch and Didion, 1993).

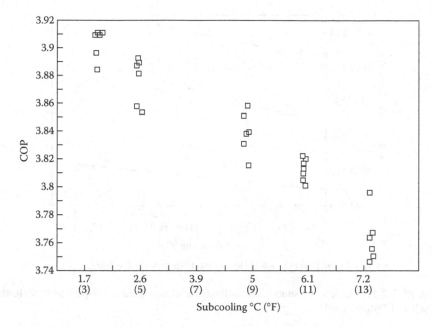

FIGURE 5.2.11 Heating mode optimization for R22; *x*-flow: crossflow, *x*-para: cross-parallel flow (Rothfleisch and Didion, 1993).

In summary, there are two competing effects to be balanced: a positive effect on the COP by a larger latent heat and a negative effect by an increased condensing pressure.

One important characteristic of the R32/R134a mixture obtained from these figures is that the capacity of R32/R134a increases slower than that of R22 while the power increases faster than that of R22 when the subcooling increases. This fact means the optimum COP of R32/R134a occurs at a lower subcooling than that of R22. This feature can be understood from the vapor pressure curve of these two refrigerants. R32/R134a has a greater slope than that of R22; therefore, the greater slope of R32/R134a causes a faster condensing pressure increase than R22 at the same temperature increase.

Since the retrofit approach was used in this study, the heat exchanger configuration was the same for the two refrigerants but it changed from crossflow (referred to as *x*-flow) to cross-parallel flow (referred to as *x*-para) when the operating mode switched from cooling to heating. There was a definite optimum point for R22 cooling optimization results but no optimum point for the R22 heating optimization results, as shown in Figures 5.2.10 and 5.2.11.

Since these tests were done for the same R22, the different results are believed due to the difference in the condenser configuration. In fact, with help from Figures 5.2.12 through 5.2.14, one can explain this phenomenon. The outlet refrigerant of the cross-parallel flow heat exchanger will approach the exiting air temperature while that of the crossflow heat exchanger will approach the entering air temperature.

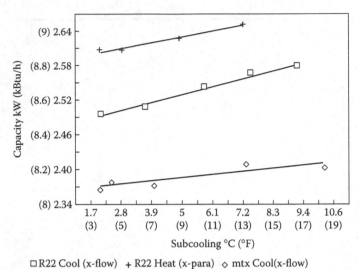

FIGURE 5.2.12 Capacity vs. subcooling; *x*-flow: crossflow, *x*-para: cross-parallel flow (Roth-fleisch and Didion, 1993).

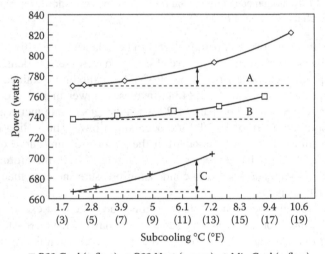

FIGURE 5.2.13 Power vs. subcooling; *x*-flow: crossflow, *x*-para: cross-parallel flow (Roth-fleisch and Didion, 1993).

Therefore, the condenser outlet refrigerant temperature in the heating mode should be higher than that in the cooling mode since the approach temperature cannot be negative. The condensing temperature in the heating mode should then be higher than that in the cooling mode when the same degree of subcooling is maintained. This reason is why the power and condensing pressure in the heating mode increases faster than in the cooling mode.

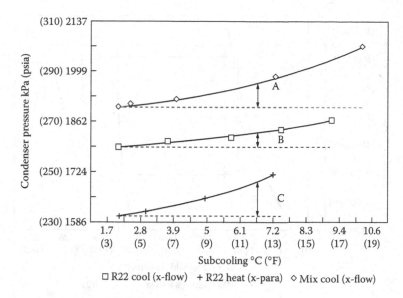

FIGURE 5.2.14 Condensing pressure vs. subcooling (Rothfleisch and Didion, 1993).

TABLE 5.22
Retrofit Performance of R32/R134a as compared to R22

Test Condition [°C]	Normalized Capacity	Normalized COP	Normalized Power
Cooling test A [35]	0.93	0.89	1.04
Cooling test B [27.8]	0.94	0.90	1.05
Heating test 47S [8.3]	0.93	0.91	1.02
Heating test 17L [–8.3]	1.04	0.90	1.16

Source: Rothfleisch and Didion, 1993.

The retrofit performance data of R32/R134a normalized by R22 results for the same test are summarized in Table 5.22. This table shows the lower capacity and COP by about 6 to 10% but the heating capacity increases 4% at the –8.3°C heating test due to the passive mass fraction shift.

One of the reasons for the lower performance of R32/R134a is due to the lower volumetric capacity of R32/R134a. The latent heat of R32/R134a was 17% higher than that of R22 but the suction density was 21% lower than that of R22 for the cooling Test A condition, which resulted in an 8% lower volumetric capacity. Figure 5.2.15 compares two refrigerant cycles in the $\ln(P)$-h diagram. From Figure 5.2.15, a higher condensing pressure and larger latent heat of evaporation of R32/R134a than those of R22 can be easily recognized.

SLHX Test of R32/R134a

To investigate the cycle options for the R32/R134a mixture performance improvement, further tests with SLHX were conducted in the cooling mode. The low-pressure

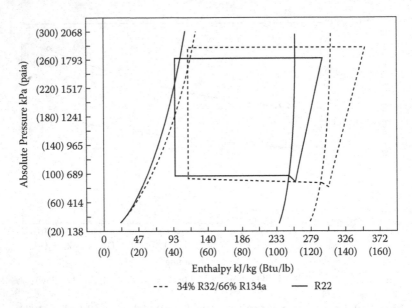

FIGURE 5.2.15 *P-h* diagram of R22 and R32/R134a (34/66 wt.%) (Rothfleisch and Didion, 1993).

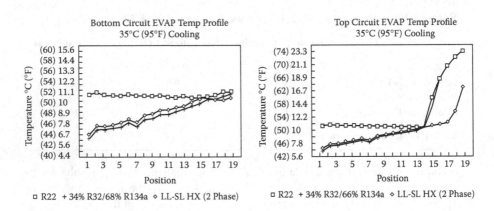

FIGURE 5.2.16 Evaporator temperature profile of R22 and R32/R134a (34/66 wt.%). (a) Bottom circuit. (b) top circuit. (Rothfleisch and Didion, 1993).

inlet side of the SLHX can be either single-phase or two-phase by adjusting the expansion valve. Both test results at Test A and B conditions are summarized in Table 5.23. From the test results, the SLHX cycle with optimized expansion valve opening at Test A condition showed a performance degradation at Test B condition, even worse than without SLHX.

For the cycle without SLHX, the refrigerant migrates from the evaporator to the condenser when the ambient temperature drops since the compressor delivers more

TABLE 5.23
SLHX Performance of R32/R134a as compared to R22

Test Condition [°C]	Refrigerant Phase Entering Low-pressure Side of SLHX	Normalized Capacity	Normalized COP	Normalized Power
Cooling test A [35]	No SLHX	0.93	0.89	1.04
	1 phase	0.95	0.90	1.05
	2 phase	0.96	0.91	1.06
	2 phase[a]	0.95	0.90	1.06
Cooling test B [27.8]	No SLHX	0.94	0.90	1.05
	1 phase	0.91	0.87	1.04
	2 phase	0.92	0.88	1.04
	2 phase[a]	0.97	0.92	1.05

Note: The expansion valve was adjusted at the 35°C test condition.

[a] The expansion valve was adjusted at the 27.8°C test condition.

Source: Rothfleisch and Didion, 1993.

refrigerants to the condenser, while the expansion device further limits the refrigerant flow due to the reduced pressure ratio. For the cycle with SLHX, the refrigerant flow through the expansion valve is limited due to the increased subcooling by the SLHX, which contributes to the lower performance. When the expansion valves were adjusted at Test B (27.8°C) condition, the performance was improved for Test B while Test A had an identical performance.

Conclusions

- The heating optimization should be conducted while balancing the heating COP at the high ambient temperature and the heating capacity at the low ambient temperature.
- The compressor power increases as the degree of subcooling increases. The capacity of R32/R134a increases slower than that of R22, while the power increases faster than that of R22 when the subcooling increases. This fact means the optimum COP of R32/R134a occurs at lower subcooling than that of R22.
- The retrofit performance of R32/R134a showed the lower capacity and COP than those of R22 by about 6 to 10%. The heating capacity increased by 4% at the −8.3°C heating test due to the passive mass fraction shift.
- The SLHX cycle with optimized expansion valve opening at Test A condition showed a performance degradation at Test B condition, even worse than without SLHX. When the expansion valves were adjusted at Test B (27.8°C) condition, the performance was improved for Test B while Test A had an identical performance.

REFERENCES

Air-Conditioning and Refrigeration Institute, 1994, "Unitary Air-Conditioning and Air Source Heat Pump Equipment," ARI Standard 210/240.

Allied Signal Chemicals, 1995, Genetron AZ-20 Product Brochure.

American Society of Heating, Refrigerating and Air-Conditioning Engineers, Inc., 1995, "Methods of Testing for Seasonal Efficiency of Unitary Air-conditioners and Heat Pumps," ASHRAE STANDARD ANSI/ASHRAE 116.

ASHRAE Handbook of Fundamentals, 1985, Chapter 16, "Refrigerants," American Society of Heating, Refrigeration and Air Conditioning Engineers, Atlanta, GA.

Calm J.M., 1995, "Refrigerants and Lubricants Data for Screening and Application," Proceedings of the 1995 International CFC and Halon Alternatives Conference, Washington D.C., pp. 169–178.

Conklin, J.C. and E.A. Vineyard, 1991, "Cycle Performance Comparison between a Non-azeotropic Refrigerant Mixture and R22," Proceedings of XVIIIth International Congress of Refrigeration, Vol. III, pp. 1270–1274.

Didion, D.A. and D. Bivens, 1989, "The Role of Refrigerant Mixtures as Alternatives," CFC Technical Conference '89 at the National Institute of Standards and Technology, American Society of Heating, Refrigeration and Air Conditioning Engineers, Atlanta, GA.

Domanski, P.A. and M.O. McLinden, 1992, "A Simplified Cycle Simulation Model for the Performance Rating of Refrigerants and Refrigerant Mixtures," *International Journal of Refrigeration*, Vol. 15, No. 2.

DuPont, 1994, Suva Refrigerants Technical Information.

DuPont, 1995, Suva Alternative Refrigerants Technical Information.

Gallagher, J., M. McLinden, and M. Huber, 1993, REFPROP NIST Thermodynamic Properties of Refrigerants and Refrigerant Mixtures Database, Version 4.01, Thermophysics Division of National Institute of Standards and Technology, Gaithersburg, MD.

Godwin, D., 1993, "Results of System Drop-In Tests in ARI's R-22 Alternative Refrigerants Evaluation Program," ARI, Arlington, VA.

Hwang, Y., J. Judge, and R. Radermacher, 1995, "An Experimental Evaluation of Medium and High Pressure HFC Replacements for R22," Proceedings of the 1995 International CFC and Halon Alternatives Conference, Washington D.C., pp. 41–48.

Kauffeld, M., W. Mulroy, M. McLinden, and D. Didion, 1990, "An Experimental Evaluation of Two Nonazeotropic Refrigerant Mixtures in a Water-To-Water, Breadboard Heat Pump," NISTIR 90-4290, National Institute of Standards and Technology, Gaithersburg, MD.

Kolliopoulos, K.G., 1994, "Update on Developmental HFC Replacements for HCFC-22," Presented at the 1994 International CFC and Halon Alternatives Conference, Washington, D.C.

Kruse, H., 1980, "The Advantages of Non-Azeotorpic Refrigerant Mixtures for Heat Pump Application," Mons (Belgium) Meeting of Commissions D1, D2, E1 and E2 of the IIR.

McLinden, M. and R. Radermacher, 1987, "Methods for Comparing the Performance of Pure and Mixed Refrigerants in the Vapor Compression Cycle," *International Journal of Refrigeration*, Vol. 10, No. 6.

Morrison, G. and M. McLinden, 1986, "Application of a Hard Sphere Equation of State to Refrigerants and Refrigerant Mixtures," NBS Technical Note 1226, National Institute of Standards and Technology, Gaithersburg, MD.

Mulroy, W. and D.A. Didion, 1986, "The Performance of a Conventional Residential Sized Heat Pump Operating with a Nonazeotropic Binary Refrigerant Mixture," NBSIR 86-3422, National Institute of Standards and Technology, Gaithersburg, MD.

Mulroy, W., M. Kauffeld, M. McLinden, and D.A. Didion, 1988, "An Evaluation of Two Refrigerant Mixtures in a Breadboard Air Conditioner," Proceedings of the IIR Conference at Purdue, West Lafayette, IN.

Mulroy, W., M. Kauffeld, M. McLinden and D.A. Didion, 1988, "Experimental Evaluation of Two Refrigerant Mixtures in a Breadboard Air Conditioner," Proceedings of the 2nd DOE/ORNL Heat Pump Conference, CONF-8804100.

Pannock, J. and D.A. Didion, 1991, "The Performance of Chlorine-Free Binary Zeotropic Refrigerant Mixtures in a Heat Pump," NISTIR 4748, National Institute of Standards and Technology, Gaithersburg, MD.

Quast, U. and H. Kruse, 1986, "Experimental Performance Analysis of a Reciprocating Compressor Working with Non-Azeotropic Refrigerant Mixtures," Proceedings of the International Compressor Engineering Conference at Purdue, West Lafayette, IN.

Radermacher, R., 1986, "Advanced Versions of Heat Pump with Zeotropic Refrigerant Mixtures," *ASHRAE Transactions*, Vol. 92, Pt. 2, pp. 52–59.

Rothfleisch, P.I. and D.A. Didion, 1991, "A Performance Evaluation of a Variable Speed, Mixed Refrigerant Heat Pump," NISTIR 93-5321, National Institute of Standards and Technology, Gaithersburg, MD.

Rothfleisch, P.I. and D.A. Didion, 1993, "A Study of Heat Pump Performance Using Mixtures of R32/R134a and R32/R125/R134a as Drop-In Working Fluids for R22 with and without a Liquid-Suction Heat Exchanger, NISTIR 5321, National Institute of Standards and Technology, Gaithersburg, MD.

Vakil, H., 1981, "Means and Method for the Recovery of Expansion Work in a Vapor Compression Cycle Device," United States Patent No. 4034099.

Vineyard, E.A, 1988, "Laboratory Testing of a Heat Pump System Using an R13B1/R152a Refrigerant Mixture," *ASHRAE Transactions*, Vol. 94, Pt. 1, pp. 292–303.

Vineyard, E.A. and J.C. Conklin, 1991, "Cycle Performance Comparison between a Nonazeotropic Refrigerant Mixture and R22," Proceedings of XVIIIth International Congress of Refrigeration, Vol. II, pp. 543–547.

Vineyard, E.A. and J.C. Conklin, 1993, "Cycle Performance Testing of Nonazeotropic Mixtures of HFC32/HFC124 with Enhanced Surface Heat Exchangers," *ASHRAE Transactions*, Vol. 99, Pt. 2, pp. 97–103.

Vineyard, E.A., J.C. Conklin, and A.J. Brown, 1993, "Cycle Performance Testing of Nonazeotropic Mixtures of HFC-143a/HCFC-124 and HFC32/HCFC124 with Enhanced Surface Heat Exchangers, *ASHRAE Transactions,* Vol. 99, Pt. 2, pp. 97–103.

6 Refrigerant Mixtures in Refrigeration Applications

As the global economy and technology have improved dramatically since World War II, the demand for human comfort has resulted in an increased demand for household refrigerators and freezers. Energy usage has increased also by the increased use of energy-consuming products. This change of lifestyle generates the demand to provide additional refrigeration capacity with increased energy efficiency. The introduction of refrigerant mixtures was considered mainly due to the increased demand of the larger refrigeration capacity, the better control of two evaporator temperature levels and the potential of improved energy efficiency. Refrigerators with two temperature levels were considered good applications for refrigerant mixtures because of one main inherent thermodynamic irreversibility: in conventional refrigerators and freezers, the higher temperature food compartment is cooled by an evaporator operating on the freezer temperature level.

6.1 SINGLE EVAPORATOR REFRIGERATION CYCLE

Refrigerators or freezers with single evaporator and one temperature level compartment (SEOC) utilize the refrigeration cycle as shown in Figure 3.3.1. A *T-s* diagram of the SEOC· is shown in Figure 3.4.1(a). The dotted line (1A–4A) is the air temperature of the refrigerating or freezing compartment as it varies along the length of the evaporator. The dotted line (3A–2A) is the air temperature across the condenser as it varies along the length of the condenser.

Most of household refrigeration machines are a combination of refrigerators and freezers used to maintain two low temperature levels. The lowest temperature level compartment is called the "freezer compartment" and the other compartment has a higher temperature and is called the "food compartment." Typical compartment temperatures are about −15°C for the freezer compartment and about 3°C for the food compartment. Refrigerator and freezers with a single evaporator and two temperature level compartments (SETC) utilize the same refrigeration cycle as that of SEOC. Conventional refrigerators and freezers with a single evaporator then employ a damper to control the degree of heat exchange between the compartments, as shown in Figure 6.1.1. A *T-s* diagram of SETC is shown in Figure 6.1.2. Dotted lines (1A–4′A and 3A–2′A) are air temperatures across the evaporator and condenser, respectively, which vary along the length of the heat exchangers for counter-current heat exchange.

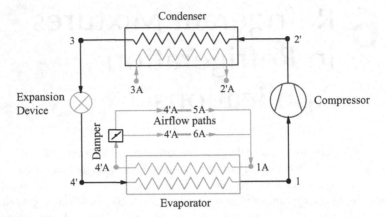

FIGURE 6.1.1 Single evaporator refrigeration cycle with two temperature level compartments.

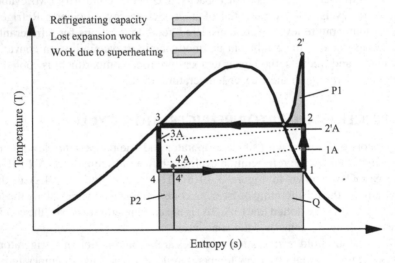

FIGURE 6.1.2 *T-s* diagram of single evaporator refrigeration cycle with two temperature level compartments.

To maintain the temperature in the freezer compartment at the desired level, the low evaporation temperature must also be used to cool the food compartment. Thus, the *temperature lift*, the temperature difference between the condenser and the evaporator, is necessarily greater for the SETC than for the appliances with SEOC.

The inherent problem with SETC is that it does not take advantage of the smaller temperature lift available for the food compartment, resulting in a lower overall COP. The main reasons for employing the SETC concept are space, cost savings and excellent reliability. Since the Montreal Protocol declared a ban on CFCs (chlorofluorocarbons) used as refrigerants and foaming agents of refrigeration systems, there has been an effort to find CFC substitutes for the single evaporator refrigeration cycle.

TABLE 6.1

Simulation Results of Refrigerants Examined

Pure Refrigerant	COP	Volumetric Capacity [kg/m³]	Mixed Refrigerant	COP	Volumetric Capacity [kg/m³]
R11	1.15	49	R22/152a	1.37	851
R12	1.35	769	R22/142b	1.39	706
R13B1	1.28	1964	R22/134a	1.35	1115
R22	1.35	1247	R22/134	1.36	839
R32	1.34	2047	R22/123	1.31	168
R114	1.25	182	R32/152a	1.38	1001
R123	1.15	42	R32/142b	1.41	544
R124	1.32	406	R32/134a	1.35	1170
R125	1.24	1580	R32/134	1.36	862
R134	1.33	545	R125/152a	1.36	671
R134a	1.33	715	R125/142b	1.39	499
R141b	1.35	77	R125/141b	1.40	127
R142b	1.36	367	R125/134a	1.33	715
R143a	1.29	1464	R125/134	1.34	670
R152a	1.36	671	R125/123	1.28	100
			R143a/152a	1.36	671
			R143a/142b	1.39	528
			R143a/141b	1.41	140
			R143a/134a	1.33	777
			R143a/134	1.34	694
			R143a/123	1.29	116

Source: Jung and Radermacher, 1991.

Stocker and Walukas simulated the performance of the SETC with R12/R114 mixtures (1981). In their model, the condenser outlet and the evaporator outlet were assumed to be a saturated liquid and vapor, respectively. In their study, they showed when the mass fraction of R114 is 0.5, the power consumption is 11% less than when the mass fraction of R114 is 0.01.

As a preliminary study to screen potential alternative refrigerants, Jung and Radermacher (1991) presented their simulation results for 15 pure and 21 mixed refrigerants. A list of refrigerants examined with results are shown in Table 6.1. Based on these results, they concluded that there is no drop-in pure refrigerant because R134, R134a and R152a have a 7 to 30% lower capacity than R12. Moreover, R22, R32, R152a and R142b have comparable COPs to R12 but have far different capacities than R12. As drop-in refrigerants, they recommended R22/R142b, R22/R152a and R32/R142b.

In an experimental effort, Vineyard et al. (1989) investigated the energy-saving potential of five refrigerants: R12, R500, R12/dimethylether (DME), R22/R142b and R134a. From their drop-in tests for these five refrigerants in a 0.5 m³ top-mount refrigerator-freezer, they found that azeotropic mixtures, R500 and R12/DME saved about 6% on energy consumption, while R134a and R22/R142b showed about 8% higher energy consumption.

Zhang et al. (1992) investigated the performance potential of a ternary mixture, R22/R114/R152a (36/40/24 wt.%). The motivation of this study was to reduce the use of CFCs. Zhang et al. also experimentally compared the performance of R152a with R12 while keeping the structure of the refrigerator and only optimizing the charge for each refrigerant. They found that R152a could reduce the energy consumption by 10% over R12; however, they also found that the evaporating temperature is about 7 K higher than that of R12. To match the evaporating temperature as a retrofit refrigerant of R12 and decrease the flammability of R152a, Zhang et al. tried to mix R152a with R22. They found that R22/R152a has a lower evaporating temperature than R152a but is still 4 K higher than that of R12 but the energy consumption of R22/R152a is 2% higher than R12. Therefore, R22/R152a does not have any advantage in terms of energy savings. They then investigated R22/R114/R152a to take advantage of R152a's properties and avoid any flammability issues of R152a by mixing it with R22 and R114. They found that the energy consumption of R22/R114/R152a mixture is 3% lower than that of R12.

The Montreal Protocol aims not just at reducing the use of CFCs but banning their use entirely. Therefore, the R22/R114/R152a mixture cannot be an ultimate solution because it contains R114. For a solution to this problem, Kuijpers et al. (1990) investigated the performance potential of a ternary mixture, R22/R124/R152a (28/40/32 wt.%). Another motivation of Kuijpers' study was to avoid the perceived energy penalty of R134a, regarded as one of the alternatives of R12. They found that the COP of R22/R124/R152a mixture is 6% higher than that of R12 when the evaporation temperature is between −15 and −25°C.

6.2 DUAL EVAPORATOR REFRIGERATION CYCLE

Refrigerator-freezers with dual evaporators and two temperature level compartments (DETC) utilize the refrigeration cycle as shown in Figure 6.2.1. These devices have separate compartments, each fitted with an evaporator: the freezer compartment with

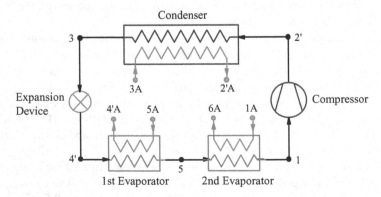

FIGURE 6.2.1 Dual evaporator refrigeration cycle with two temperature level compartments.

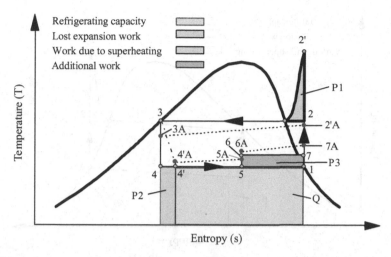

FIGURE 6.2.2 *T-s* diagram of dual evaporator refrigeration cycle with two temperature level compartments.

the first evaporator and the food compartment with the second evaporator. A *T-s* diagram of SETC is shown in Figure 6.2.2. Dotted lines (5A–4'A and 7A–6A) are air temperatures of the freezer compartment and the food compartment, respectively, that vary along the length of the first and second evaporators. The dotted line (3A–2'A) is the air temperature along the condenser. The requirement of having two separate compartment temperatures results in an additional loss for dual-evaporator refrigeration systems as clearly represented by the area P3. As for the SETC, the lower evaporation temperature must be used for the food compartment also in order to maintain the lower temperature in the freezer compartment. The inherent problem with DETC is the same as that of SETC, resulting in a low overall COP.

6.3 LORENZ-MEUTZNER CYCLE

There has been an effort to improve the disadvantage of the single evaporator refrigeration cycle by implementing the Lorenz refrigeration cycle using refrigerant mixtures. Figure 6.3.1 shows the *T-s* diagram of the Lorenz cycle with DETC. This figure shows the same benefit of the Lorenz refrigeration cycle over the Rankine refrigeration cycle for SEOC.

The second major effort to improve the disadvantage of the single evaporator refrigeration cycle uses the dual-evaporator refrigeration cycle. Lorenz and Meutzner (1975) proposed a dual evaporator refrigeration cycle with two inter-coolers using zeotropic refrigerant mixtures, as shown in Figure 6.3.2. In this cycle, the condenser outlet is connected to the suction heat exchanger that exchanges heat with the suction line. The liquid outlet of the suction heat exchanger is connected to an internal heat exchanger that subcools the liquid heater with the evaporating refrigerant as it passes between the evaporators. The outlet of the internal heat exchanger is connected to the expansion device.

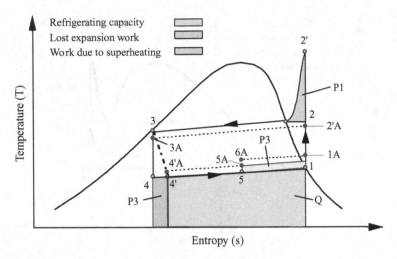

FIGURE 6.3.1 *T-s* diagram of dual evaporator Lorenz cycle with two temperature level compartments.

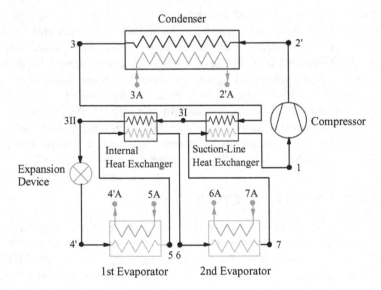

FIGURE 6.3.2 Lorenz-Meutzner cycle.

The expanded two-phase refrigerant from the expansion device enters the evaporator and then passes through the internal heat exchanger, the food temperature evaporator and the suction heat exchanger, sequentially. Therefore, the internal heat exchanger and suction line heat exchanger offer more subcooling at the condenser outlet and at the same time take advantage of and provide an appropriate temperature glide for both evaporators. Figure 6.3.3 shows the *T-s* diagram of the Lorenz-Meutzner cycle. Here the additional work requirement, area P3 of Figure 6.3.1, is not present because the refrigerant entering the refrigerator evaporator is heated by

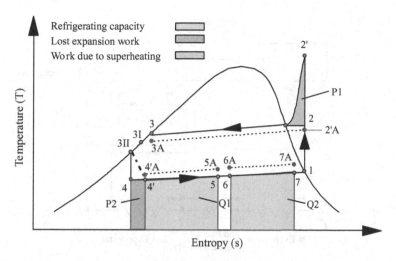

FIGURE 6.3.3 *T-s* diagram of a Lorenz-Meutzner cycle.

the second heat exchanger from T_5 to T_6. Therefore, an approach temperature between the air and refrigerant of the second evaporator is reduced. With the addition of a food compartment evaporator and the use of a well-chosen zeotropic refrigerant mixture, evaporation at two temperature levels in a single refrigeration cycle can be achieved with a higher COP than the conventional vapor compression refrigeration cycle. This is the concept upon which the Lorenz-Meutzner cycle is based.

Lorenz and Meutzner reported a 20% energy savings for a Lorenz-Meutzner cycle with a mixture of R22/R11 (50/50 wt.%) compared to a conventional refrigerator with R12. Stoecker and Walukas (1981) showed no energy savings for a mixture of R12/R114 (50/50 wt.%) but they reported that the simulated power of the Lorenz-Meutzner cycle with the R12/R114 mixture (50/50 wt.%) was 12% less than that of R12. In their study, they showed that when the mass fraction of R114 is 0.5, the power consumption is 11% less than when the mass fraction of R114 is 0.01. Tiedemann et al. (1991) tested various mixtures but did not find any savings. The experimental results of Rose et al. (1992) showed a 9% reduction in energy consumption with a zeotropic mixture of R22/R123. Sand et al. (1993) found that the Lorenz-Meutzner cycle with R32/R124 mixtures (15/85 wt.%) saved 6 to 7% energy use over R12. They also emphasized the importance of counterflow, refrigerant-to-air heat exchangers.

6.4 MODIFIED LORENZ-MEUTZNER CYCLE REFRIGERATOR

The Lorenz-Meutzner cycle utilizes two internal heat exchangers to reduce the approach temperature between the refrigerant and air streams at the freezer and food compartments by subcooling the liquid refrigerant before the expansion process. Radermacher and Jung (1993a) presented a modified version of the Lorenz-Meutzner cycle that further subcools the liquid refrigerant before the expansion process by extending the high pressure liquid line through the first and second evaporators, as

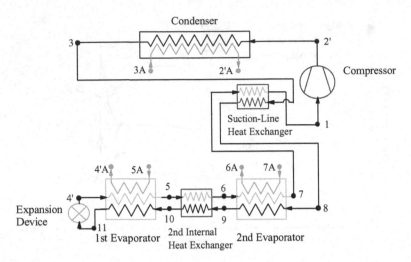

FIGURE 6.4.1 Modified Lorenz-Meutzner cycle.

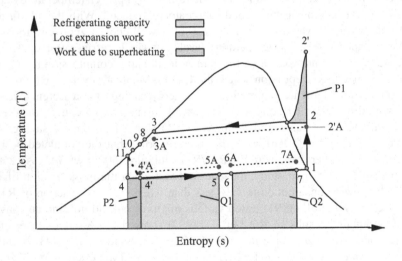

FIGURE 6.4.2 *T-s* diagram of a modified Lorenz-Meutzner cycle.

shown in Figure 6.4.1. This is an example for implementing the three paths evaporator design that is discussed in Chapter 4. The difference between the Lorenz-Meutzner cycle and the modified Lorenz-Meutzner cycle is shown in the *T-s* diagram of the modified Lorenz-Meutzner cycle refrigerator, Figure 6.4.2. The modified Lorenz-Meutzner cycle refrigerator has a four step subcooling process, while the Lorenz-Meutzner cycle refrigerator has only two steps of subcooling. In the modified Lorenz-Meutzner cycle, the first and second evaporators serve as three-way heat exchangers made by a smaller tube (liquid line) inserted into a larger tube (suction line and evaporator). The compartment air surrounding the bigger tube and subcooled liquid in the smaller tube rejects heat to the two-phase refrigerant running counterflow through the annulus of the two tubes. Zhou et al. (1994) reported 16.6, 14.6 and 16.7% energy savings from their experiments with a modified-Lorenz-Meutzner cycle refrigerator with zeotropic mixtures R22/R123, R290/n-pentane and R290/R600, respectively. The conventional SETC uses a damper to control the food compartment temperature and a thermostat to control the freezer compartment temperature. The main purpose of the dual evaporators is to provide better temperature control.

For the Lorenz-Meutzner cycle, however, independent temperature control of the two compartments is not as simple since there is no damper. The temperature conditions of both compartments compared to their temperature settings determine the operation of the compressor. Therefore, one compartment temperature can be dropped further than the set temperature when the other compartment temperature is higher than the temperature set point. In this case, failure of implementing independent temperature control results in an energy loss. Simmons et al. (1995) suggested an independent temperature control of compartments in a modified Lorenz-Meutzner cycle by using a bistable solenoid valve as shown in Figure 6.4.3. The suggested control of a modified Lorenz-Meutzner cycle directs refrigerant through alternate evaporator paths with a bistable solenoid valve that regulates the entire flow of refrigerant to only the freezer evaporator or to both evaporators.

When the bistable solenoid valve is at position A (Mode A), the cycle becomes a modified Lorenz-Meutzner cycle as shown in Figure 6.4.3. When the bistable solenoid valve is at position B (Mode B), the cycle becomes SEOC with two serial suction line heat exchangers. When both compartments need cooling, the bistable valve switches to Mode A. When only the freezer compartment needs cooling, the bistable valve switches to Mode B. Therefore, the suggested control method for a modified Lorenz-Meutzner cycle provides a similar independent temperature control to the SETC with the damper.

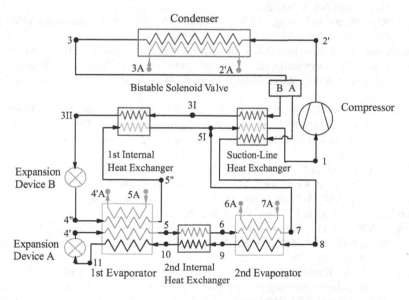

FIGURE 6.4.3 Modified Lorenz-Meutzner cycle with bistable solenoid valve.

Simmons et al. (1995) experimentally demonstrated that the modified Lorenz-Meutzner cycle with a bistable solenoid valve provides the desired level of temperature control in the compartments. They also reported that an energy penalty of 6% resulted due to the addition of the control valve and the resulting refrigerant charge increase of 17% over the optimized baseline unit. These are 17% net savings over a system with pure refrigerants.

REFERENCES

Ashley, C., 1945, U.S. Patent No. 2492725.

Dmitriev, W. et al., 1974, "Comparative Tests of Hermetic Compressor FR-O.15 on R12 and Non-azeotropic Mixture of R12/R143," *Cholodilnaja Technika*, Vol. 51, No. 6, pp. 25–27.

Jung, D.S. and R. Radermacher, 1991, "Simulation of a Single-Evaporator Refrigerator," *International Journal of Refrigeration*, Vol. 14, pp. 223–232.

Kuijpers, L.J.M., J.A. de Wit, A.A.J. Benschop and M.J.P. Janssen, 1990, "Experimental Investigations into the Ternary Blend HCFC-22/124/152A as a Substitute in Domestic Refrigeration," IIR Purdue Refrigeration Conference.

Liu, Z., I. Haider, and R. Radermacher, 1995, "Simulation and Test Results of Hydrocarbon Mixtures in a Modified-Lorenz-Meutzner Cycle Domestic Refrigerator," *HVAC&R Research*, Vol. 1, No. 2, pp. 127–142.

Lorenz, A. and K. Meutzner, 1975, "On application of non-azeotropic two component refrigerants in domestic refrigerators and home freezers," XIV International Congress of Refrigeration, Moscow.

Phillipp, L., 1948, U.S. Patent No. 2483842.

Radermacher, R. and D.S. Jung, 1992, "Refrigeration System," U.S. Patent No. 5,092,138.

Radermacher, R. and D.S. Jung, 1993a, "Refrigerator System for Two-Compartment Cooling," U.S. Patent No. 5,235,820.

Radermacher, R. and D.S. Jung, 1993b, "Subcooling System for Refrigeration Cycle," U.S. Patent No. 5,243,837.

Radermacher, R., H. Ross and D.A. Didion, 1983, "Experimental Determination of Forced Convection Evaporative Heat Transfer Coefficients for Non-Azeotropic Refrigerant Mixtures," ASME Winter Annual Meeting, Boston, MA, Paper No 83-WA/HT-54, Nov., pp. 13–18.

Radermacher, R., K. Kim, W. Kopko and J. Pannock, 1995, U.S. Patent No. 5,406,805.

Rose, B., D.S. Jung and R. Radermacher, 1992, "Testing of Domestic Two-evaporator Refrigerators with Zeotropic Refrigerant Mixtures," *ASHRAE Transactions*, Vol. 98, Pt. 2, pp. 216–226.

Sand, J.R., E.A. Vineyard and V.D. Baxter, 1993, "Laboratory Evaluation of An Ozone-Safe Nonazeotropic Refrigerant Mixture in a Lorenz-Meutzner Refrigerator Freezer Design," *ASHRAE Transactions*, Vol. 93, Pt. 1, pp. 1467–1481.

Schubert, A. and Herrick, C., 1964, U.S. Patent No. 3,019,614.

Stoecker, W.F. and D.J. Walukas, 1981, "Conserving Energy in Domestic Refrigerators through the Use of Refrigerant Mixtures," *International Journal of Refrigeration*, Vol. 4, pp. 201–208.

Tiedemann, T., M. Kauffeld, K. Beermann and H. Kruse, 1991, "Evaluation of ozone-safe, low greenhouse warming potential zeotropic refrigerant mixtures in household refrigerators/freezers," XVIIIth International Congress of Refrigeration, Montreal, Canada.

Tiedemann, T. and H. Kruse, 1994, "Evaluation of zeotropic hydrocarbon mixtures in a Lorenz-Meutzner-cycle refrigerator-freezer," New Applications of Natural Working Fluids in Refrigeration and Air Conditioning, IIR Conference, Hannover, Germany.

Vineyard, E.A., J.R. Sand and W.A. Miller, 1989, "Refrigerator-Freezer Energy Testing with Alternative Refrigerants," *ASHRAE Transactions*, Vol. 95, Pt. 2, pp. 295–299.

Vineyard, E.A., L. Roke and F. Hallett, 1991, "Overview of CFC Replacement Issues for Household Refrigeration," Proceedings of 1991 International CFC and Halon Alternative Conference, Washington D.C., pp. 310–316.

Zhou, Q., J. Pannock and R. Radermacher, 1994, "Development and Testing of a High Efficiency Refrigerator," *ASHRAE Transactions*, Vol. 100, Pt. 1, pp. 1351–1358.

Zhou, Q. and R. Radermacher, 1997, "Development of a Vapor Compression Cycle with a Solution Circuit and Desorber/Absorber Heat Exchanger," *International Journal of Refrigeration*, Vol. 20, No. 2, pp. 85–95.

7 Refrigerant Mixtures in Heat Pump Applications

Refrigerant mixtures were introduced for heat pump applications primarily to meet environmental concerns. However, the use of refrigerant mixtures was considered as early as the late 1800s and then again in the 1950s and thereafter to improve their capacity control and energy efficiency. The use of mixtures was studied as alternatives of hydrochlorofluorocarbons (HCFCs) after the inclusion of HCFCs as one of controlled substances in the Montreal Protocol in 1992.

7.1 CAPACITY CONTROL

Heat pumps must be operated over the wide range of ambient temperatures. Furthermore, the cooling and heating capacity of heat pumps tends to vary in opposition to the load the heat pump is intended to match. The cooling capacity, for example, decreases but the cooling load increases as the ambient temperature rises. The heating capacity of the heat pump decreases but the heating load increases as the ambient temperature drops. When the temperature difference between the ambient temperature and the desired conditioned space temperature is larger than the system design condition, cooling or heating loads are larger than the cooling or heating capacity of heat pumps. In this case, the capacity of the heat pumps is smaller than the cooling or heating load though the heat pumps are operated with their full capacity. When the temperature difference between the ambient temperature and the desired conditioned temperature is smaller than the system design condition, cooling or heating loads are smaller than the cooling or heating capacity of the heat pumps. In this case, the heat pumps are only operated with partial capacity.

To operate the heat pumps with partial capacity, three general methods have been utilized: cyclic operation, variable speed compressors using inverter control and variable displacement compressors. Cyclic operation can have the disadvantage of energy loss due to the excessive power consumption during the start up period. Variable speed compressors have a disadvantage of higher cost and complexity of system design, though they have an advantage of flexibility of control. Variable displacement compressors are limited to automotive applications or to large systems when individual cylinders are unloaded.

Refrigerant mixtures provide one more option for capacity control. If one component of a mixture is separated from the circulating working fluid mixture and stored selectively away from the cycle, the circulating mixture mass fraction is affected by the amount of stored component. If a binary mixture of two pure refrigerants with different boiling points is used, the mass fraction of the mixture changes during its phase change process as explained in Chapter 2. Therefore, the

221

storage of one component by exploiting the mass fraction difference in the two-phase state provides the opportunity to control the mixture mass fraction. Changes of mixture mass fraction affect the density and latent heat of the circulating mixture, resulting in capacity changes. Some options of capacity control by using refrigerant mixtures are introduced here.

Using the method of modifying mixture mass fraction to adjust capacity is most likely only suitable in a limited range around the ratio 1:2. Since the compressor speed remains unchanged, the amount of work needed just to operate the compressor (friction, oil pump and other sources) is essentially constant. Thus, during low capacity operation, the fraction of lost work increases relative to compression work done on the refrigerant and the overall system efficiency decreases.

7.1.1 ONE ACCUMULATOR CYCLE

Etherington et al. (1980) developed a variable capacity refrigeration cycle using an accumulator with a special molecular sieve that is selectively reactive to a certain refrigerant. As compared to the conventional refrigeration cycle with four basic components (evaporator, condenser, compressor and expansion device), the one accumulator cycle (OAC) has an additional accumulator as shown in Figure 7.1.1. The accumulator is installed across the high and low pressure sides and connected on one side to the condenser outlet before the expansion device and on the other side to the compressor suction after the evaporator outlet. Both sides of the accumulator can be shut off with valves C and D. Etherington et al. (1980) proposed to use a molecular sieve that only absorbs R22 in the accumulator when the cycle is charged with R12/R22.

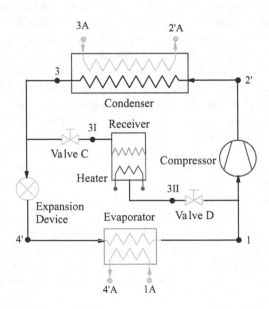

FIGURE 7.1.1 One accumulator cycle.

Assume that a binary zeotropic mixture of R12 and R22 is charged in the cycle. Normal boiling points of R12 and R22 are –29.8 and –40.8°C, respectively, and R22 has a higher saturation pressure and larger latent heat than those of R12. For example, the saturation pressures of R12 and R22 are 423 kPa and 681 kPa, respectively, at 10°C saturation temperature. The latent heats of R12 and R22 are 148 kJ/kg-K and 196 kJ/kg-K, respectively, at 10°C saturation temperature. Therefore, an R22-rich mixture works at higher operating pressures and provides more capacity. When the system capacity is larger than the load, there is a need for reducing capacity, so valve C is opened and valve D is closed. Thus, R22 is selectively absorbed into the molecular sieve. In this case, the R22 mass fraction of the circulating mixture becomes lower than the originally charged mass fraction and the system capacity is reduced due to the increase in R12 content. When the system capacity is smaller than the load, there is a need for restoring system capacity, so valve C is closed and valve D is opened. An electric heater helps to release the R22 that is absorbed in the molecular sieve. In this case, the R22 mass fraction of circulating mixture becomes the same as the originally charged mass fraction and the system capacity is restored.

A similar effect can be achieved with a single accumulator located at the evaporator outlet only. Whenever a two-phase mixture leaves the evaporator, the liquid phase is stored in the accumulator while the vapor continues to circulate through the system. The stored liquid phase is rich in the high boiler (A) and the mass fraction of the circulating mixture is shifted toward a higher content of the low boiler (B) and the system capacity is increased.

Many air conditioners that also operate as heat pumps require a suction accumulator to store excess charge during the heat pumping mode. Thus, this accumulator causes a shift in mass fraction, increasing capacity during heating only. Using only one accumulator causes a shift in mass fraction that is smaller than that possible with the two accumulator configuration.

7.1.2 TWO ACCUMULATORS CYCLE

OAC uses a selectively reactive molecular sieve to control the mass fraction of mixtures. Cycles introduced from here forward exploit the difference in boiling points of mixture components. As compared to the OAC, the two accumulators cycle (TAC) proposed by Vakil and Flock (1980) has one additional accumulator as shown in Figure 7.1.2. One accumulator is installed on the high pressure side after the condenser and the other accumulator is installed on the low pressure side after the evaporator.

Assume that a binary zeotropic mixture of a low boiling point component (A) and a high boiling point component (B) with mass fraction of x is charged in the cycle. If the high pressure accumulator (receiver) is filled with a liquid mixture and the low pressure accumulator is empty, the mass fraction of the liquid in the high pressure accumulator is the same as the originally charged mass fraction x. Therefore, the mass fraction of the circulating mixture remains the same as before and the low pressure accumulator is almost empty and allows for superheating. If the situation is reversed, the high pressure accumulator is empty and the low pressure accumulator

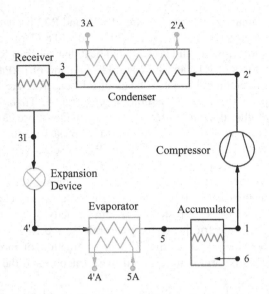

FIGURE 7.1.2 Two accumulators cycle.

is filled with a liquid mixture and the low pressure accumulator stores a mixture that is predominantly rich in component B because most of component A was already released in the evaporator. Thus, the low pressure accumulator provides a mixture rich in component A to the compressor and reduces the degree of superheating.

For example, Figures 7.1.3 and 7.1.4 show the saturated vapor density and latent heat of condensation for the binary mixtures of R22/R114. Boiling points of R22 and R114 are −40.8 and 3.8°C, respectively. Therefore, R22 works as component A and R114 works as component B. These figures show the higher mass fraction of

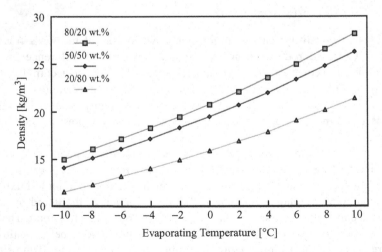

FIGURE 7.1.3 Saturated vapor density of binary mixture (R22/R114).

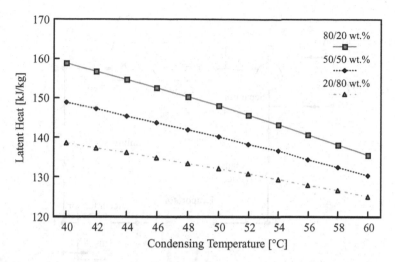

FIGURE 7.1.4 Latent heat of condensation for binary mixture (R22/R114).

R22 provides a higher saturated vapor density and latent heat of condensation at the same temperature. Therefore, the first case of filled high pressure accumulator provides smaller capacity due to the lower R22 mass fraction in the circulating mixture. The second case of a filled low pressure accumulator provides a larger capacity due to the higher R22 mass fraction in the circulating mixture. In this way, the system capacity can be adjusted by the control of mixture mass fraction. Figure 7.1.5 shows the results of capacity modulation relative to the load variation.

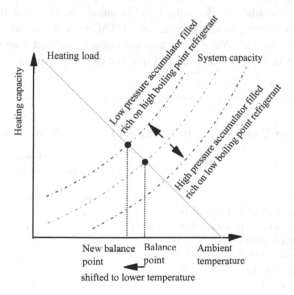

FIGURE 7.1.5 Results of capacity modulation (TAC).

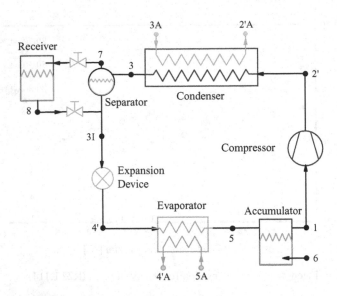

FIGURE 7.1.6 Modified two accumulators cycle.

7.1.3 MODIFIED TWO ACCUMULATORS CYCLE

To increase the capacity change range, Vakil (1979) proposed a modified two accumulators cycle (MTAC), as shown in Figure 7.1.6. In the MTAC, a separator replaces
the high pressure accumulator and is located at the condenser outlet. The high
pressure accumulator is located outside the cycle and in parallel with the separator.
Since the uncondensed vapor from the condenser is richer in component A and the
high pressure accumulator collects the liquid condensate, the mass fraction of A is
reduced in the circulating mixture. Therefore, MTAC provides a wider range of mass
fraction control than TAC. Vakil compared the capacity control ranges of these two
cycles as shown in Table 7.1. He showed that the capacity control of MTAC ranges

TABLE 7.1
Comparison of Capacity Control (R22/R114 with 50/50 mole%)

Capacity	Maximum	Minimum for TAC	Minimum for MTAC
Low accumulator	Full	Empty	Empty
R22 composition in circulating mixture [mole%]	82.2	50.0	17.6
R22 composition in low accumulator liquid [mole%]	50	13	2.5
Evaporating pressure [MPa]	0.41	0.25	0.12
Condensing pressure [MPa]	1.60	1.04	0.60
Temperature glide during condensation [K]	7.0	17.8	15.1
Heating capacity [kW]	1.00	0.67	0.38

Source: Vakil, 1979.

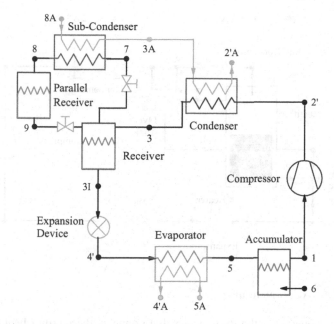

FIGURE 7.1.7 Three accumulators cycle.

from 38 to 100% using an R22/R114 binary mixture (50/50 mol.%) and from 28 to 100% using an R23/R22/R114 ternary mixture (10/60/30 mol.%).

7.1.4 THREE ACCUMULATORS CYCLE

The disadvantage of MTAC is its slow capacity modulation compared to the ambient temperature changes. To improve the capacity modulation speed, Vakil (1983) proposed a three accumulators cycle (T3AC), as shown in Figure 7.1.7. In the T3AC, a parallel accumulator is connected to the gas phase of the high pressure accumulator and is located at the condenser outlet. The inlet to the parallel accumulator passes through the subcondenser. Since the uncondensed vapor from the condenser is richer in component A, the parallel accumulator can reduce the mass fraction of A in the circulating mixture similar to the MTAC. However, T3AC can discharge the stored liquid quickly with the help of the valve next to point 9 (Figure 7.1.7). Therefore, T3AC can modulate its capacity faster than that of the MTAC.

7.1.5 CYCLE WITH RECTIFIER

As a third option to enhance the separation of a mixture, Schwind (1960) proposed a cycle with rectifier (CWR), as shown in Figure 7.1.8. In this cycle, a rectifier is installed between the condenser and expansion device. The condenser outlet is connected to the overflow tank connected to the subcondenser (5), rectifier (7) and rectifier heat exchanger (3I). A three-way valve is installed just before the expansion valve and connected to the rectifier directly (3III) and rectifier heat exchanger (3II).

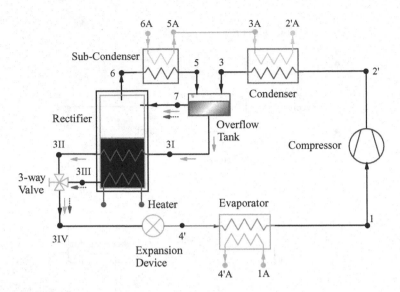

FIGURE 7.1.8 Cycle with rectifier.

In normal operation, the three-way valve connects the rectifier heat exchanger outlet (3III) and expansion valve inlet (3IV) as indicated by the dotted arrows. In this case, the overflow tank delivers liquid refrigerant from the condenser into the rectifier. The liquid refrigerant in the rectifier is fed into the expansion valve. Through this process, the mass fraction of the mixture in the rectifier and that circulating in the cycle becomes the same. If a large system capacity is required, the direction of the three-way valve is reversed and the electric heater is turned on. The refrigerant from the overflow tank (3II) is now connected to the expansion valve inlet (3IV). The low boiling point refrigerant is then vaporized and fed into the subcondenser where it is condensed again. This condensed low boiling point refrigerant enters the overflow tank and is mixed with the circulating refrigerant. Through this process, the mass fraction of low boiling point refrigerant in the circulating mixture increases and the system capacity increases. Jakobs and Kruse (1978) examined this cycle option with a binary mixture, R12/R114 (60/40 wt.%). They confirmed the circulating mass fraction of R12 changed from 50 to 90% and achieved 30% more heating capacity.

7.1.6 MODIFIED CYCLE WITH RECTIFIER

The CWR concept was commercialized in Japan by Yoshida et al. (1989) and Figure 7.1.9 shows the CWR used. In this cycle, a rectifier and reservoir are installed parallel to the expansion valve and connected by capillary tubes. The refrigerant from the rectifier to the reservoir is cooled by the compressor suction gas. Yoshida et al. used R13B1/R22 (30/70 wt.%) with boiling points of R13B1 and R22 of −57.8 and −40.8°C, respectively. During cooling (heating) at high (low) ambient temperature operation, the expansion valve is normally closed with the refrigerant sent to the reservoir. Vapor phase refrigerant rises in the rectifier and liquid from the reservoir falls into the rectifier. These two streams encounter each other in the rectifier and

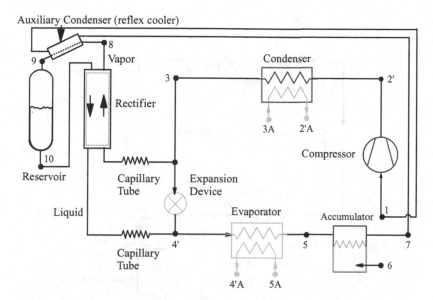

FIGURE 7.1.9 Modified cycle with rectifier.

exchange heat and mass. The high boiling point refrigerant in the rising vapor (R22) is then condensed and leaves the rectifier with the liquid stream of the rectifier's lower end. The low boiling point refrigerant in the falling liquid (R13B1) is vaporized and rises. In this way, the low boiling point refrigerant is stored in the reservoir and the high boiling point refrigerant is circulating in the cycle. Therefore, the system capacity is reduced. When a large capacity is required, the expansion valve opens and the stored low boiling point refrigerant is released from the reservoir and circulates in the cycle, resulting in a large capacity.

7.1.7 CYCLE WITH RECTIFIER-ACCUMULATOR

Rothfleisch (1995) combined the advantages of accumulator and rectifier and developed a cycle with a rectifier/accumulator (CRA), where Figure 7.1.10 shows the schematic. The rectifier/accumulator is installed between the evaporator outlet and compressor inlet and evaporator outlet and compressor inlet are connected to the top of the rectifier/accumulator. The bottom end of the rectifier is connected to the accumulator and a heater is installed in the liquid portion of the accumulator. The major difference of the CRA compared to the CWR is the location of the rectifier. The CWR has a rectifier on the high pressure side after the condenser, whereas the CRA has a rectifier on the low pressure side in the suction accumulator.

Assume a binary mixture, R32/R134a, is charged. If the heater is not in operation, the mixture is in thermal equilibrium. In this case, the rectifier/accumulator works as a conventional accumulator. If the heater is turned on, the accumulator liquid reservoir temperature is higher than the rectifier and the refrigerant from the evaporator continues to flow through the head of the rectifier. Vapor phase refrigerant from the accumulator also rises in the rectifier and liquid from the top portion of

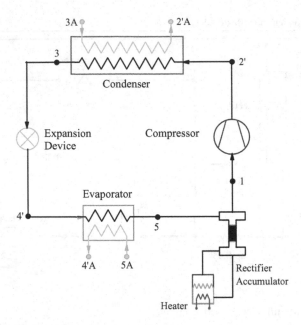

FIGURE 7.1.10 Cycle with rectifier accumulator.

the rectifier that was not released in the evaporator drops through the rectifier to the accumulator. These two streams encounter each other in the rectifier and exchange heat and mass. The high boiling point refrigerant in the rising vapor (R134a) is then condensed and leaves with the liquid. The low boiling point refrigerant in the falling liquid (R32) is vaporized and rises with the vapor. In this way, the high boiling point refrigerant (R134a) is stored in the accumulator and the low boiling point refrigerant (R32) is circulating in the cycle. The advantage of the CRA compared to the cycle with an accumulator only is a reduction in the size of the accumulator. Moreover, CRA is able to extend the composition change by eliminating the thermal equilibrium constraint.

Rothfleisch (1995) investigated the performance of the CRA with a binary mixture, R32/R134a (30/70 wt.%), and a ternary mixture, R32/R125/R134a (23/25/52 wt.%). He confirmed that the maximum mass fraction shift of the circulating mixture amounts to 54/46 wt.% for R32/R134a and 36/36/28 wt.% for R32/R125/R134a. He also found that the heating seasonal performance factor was improved by about 3% with the CRA in use.

7.2 R22 REPLACEMENT

7.2.1 SCREEN OF REFRIGERANT MIXTURES

Systematic screening approaches have been conducted to select potential refrigerant mixtures to substitute R22. Vineyard et al. (1989) applied two selecting criteria — hard criteria and soft criteria — to more than 200 pure compounds that would make up the components of zeotropic mixtures. Hard criteria (toxicity, instability and

TABLE 7.2
Performance Comparison of Pure Refrigerants Investigated by ORNL

Refrigerant	Cooling "A" Capacity [kW]	COP	Cooling "B" Capacity [kW]	COP	Heating "47S" Capacity [kW]	COP	Heating "17L" Capacity [kW]	COP
R32	17.3	2.6	20.3	3.0	15.5	3.1	8.2	2.4
R125	9.0	2.0	11.6	2.4	9.2	2.6	5.0	2.1
R143a	11.3	2.3	13.9	2.7	10.6	2.8	5.6	2.3
R22	10.1	2.6	11.8	3.0	9.6	3.0	4.7	2.4
R218	—	—	—	—	—	—	—	—
R134a	6.5	2.5	7.8	2.9	5.9	2.9	2.7	2.3
R152a	6.4	2.7	7.5	3.1	5.7	3.0	2.7	2.5
R124	3.4	2.4	4.2	2.8	2.8	2.8	1.2	2.3
R124a	—	—	—	—	—	—	—	—
R142b	3.6	2.6	4.3	3.0	3.2	3.0	1.4	2.4
C318	2.7	2.2	3.4	2.6	2.5	2.6	1.1	2.1
R143	—	—	—	—	—	—	—	—

Source: Vineyard et al., 1989.

ozone-depletion potential or ODP) were used to eliminate a refrigerant from consideration. Soft criteria (flammability, boiling point and commercial availability) were also used to eliminate a refrigerant but only as a flexible criteria. After applying these two criteria, the following seven refrigerants remained: R125, R22, R218, R134a, R124, R124a and C318. By relaxing the flammability criteria, five more refrigerants were added: R32, R143a, R152a, R142b and R143.

To evaluate performance of these twelve refrigerants at ASHRAE cooling conditions (A and B) and heating conditions (47S and 17L), the simple cycle program developed by McLinden, Cycle 5A, was utilized. Results of the cycle performance calculation are shown in Table 7.2. R218, R124a and R143 were excluded in the calculation due to the availability of property routines. From Table 7.2, R125 and C318 showed 10% lower COP than that of other refrigerants; thus, these two refrigerants were eliminated from further evaluation. Once the performance potential of pure refrigerants was evaluated, the temperature glide of the binary mixtures of selected pure refrigerants was evaluated to discover mixtures that with the best matching temperature glides with heat exchanging fluids. In this analysis, the mass fraction of each component for each mixture was fixed at 50/50 wt.%. The temperature glide of each mixture at 1102 kPa saturation pressure was then compared, as shown in Table 7.3. The authors decided the temperature glides between 10 to 30°C contained the most potential for better temperature matching in the heat exchangers, as shown with the boldface values in Table 7.3. Moreover, only R32 and R143a showed higher capacity than R22 from the cooling and heating capacity analysis shown in Table 7.2. Finally, R32/R124, R32/R142b, R143a/R124, R143a/R142b and R143a/C318 were selected as the best mixtures by combining the performance results (Table 7.2) and the temperature glides (Table 7.3).

TABLE 7.3
Temperature Glides of Binary Mixtures for 50/50 wt.%

R32	3	3	4	4	8	8	23	23	23	28	44
	R125	3	3	4	6	6	19	19	19	23	40
		R143a	3	4	5	5	17	17	17	23	38
			R22	3	4	4	11	11	11	15	29
				R218	3	3	7	7	7	10	23
					R134a	3	3	3	3	5	11
						R152a	3	3	3	4	8
							R124	3	3	3	4
								R124a	3	3	4
									R142b	3	4
										C318	3
											R143

Source: Vineyard et al., 1989.

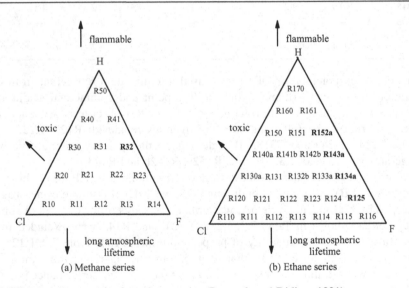

FIGURE 7.2.1 Systematization of properties (Pannock and Didion, 1991).

7.2.2 SCREEN OF HFC REFRIGERANT MIXTURES

Figure 7.2.1 shows methane- and ethane-based refrigerants arranged according to molecular structure. Pure compounds with potential as R22 alternatives are shown in Table 7.4. These six methane- and ethane-based refrigerants having zero *ozone depletion potential* (ODP) are listed and evaluated based on similar boiling points, flammability and being environmentally benign (Pannock and Didion, 1991).

Unfortunately, none of the above refrigerants satisfies all these requirements. Thus, the mixing of these refrigerants was suggested to obtain the desired properties. In 1991, the Air-conditioning and Refrigeration Institute (ARI) started the Alternative

TABLE 7.4
Characteristics of Potential Pure Component

Refrigerant	Characteristics	Normal Boiling Point [°C]
R22	Base fluid	−40.8
R23	Critical temperature is too low	−82.0
R32	Flammable, too high volumetric capacity	−51.7
R125	High DGWP, low theoretical efficiency	−48.1
R143a	Flammable, high DGWP	−47.2
R134a	Low volumetric capacity	−26.1
R152a	Flammable	−24.0

DEWP = direct global warming potential.

Refrigerants Evaluation Program (AREP) to find and evaluate promising alternatives to the refrigerants R22 and R502 for products such as unitary air conditioners, heat pumps, chillers, refrigeration equipment and ice-making machines (Godwin, 1993). This test program involved researchers worldwide and shared experimental results. AREP investigated 16 refrigerants for R22 replacements and six refrigerants for R502 replacements, as shown in Tables 7.5 and 7.6. Though no single alternative refrigerant is chosen as a universal replacement to R22, experimental results for some interesting refrigerants show a similar cooling and heating performance as R22, as shown in Table 7.7.

After years of research, three potential refrigerant mixture candidates for R22 replacements have emerged: R134a, R407C and R410A. R134a has a lower vapor pressure than R22 and results in a lower volumetric capacity. R407C has a vapor pressure similar to R22 and performs within a 6% variation. R410A has a 50% higher vapor pressure and has shown better performance within 10% variation compared to R22. In terms of vapor pressure, R407C appears to be an "easy-to-

TABLE 7.5
R22 Alternative Refrigerants Investigated by AREP

Type	Refrigerant
Pure refrigerant	R134a
	R290
	R717
Binary mixture	R32/R125 (60/40; 50/50 wt.%)
	R32/R134a (20/80; 25/75; 30/70; 40/60 wt.%)
	R125/R143a (45/55 wt.%)
Ternary mixture	R32/R125/R134a (10/70/20; 23/25/52; 24/16/60; 25/20/55; 30/10/60 wt.%)
Quaternary mixture	R32/R125/R290/R134a (20/55/5/20 wt.%)

Source: Godwin, 1993.

TABLE 7.6
R502 Alternative Refrigerants Investigated by AREP

Type	Refrigerant
Binary mixture	R125/R143a (45/55; 50/50 wt.%)
	R32/R125/R134a (10/70/20; 20/40/40 wt.%)
Ternary mixture	R32/R125/R143a (10/45/55 wt.%)
	R125/R143A/R134a (44/52/4 wt.%)

Source: Godwin, 1993.

TABLE 7.7
R22 Alternative Refrigerants Investigated by AREP

Refrigerant	Cooling Capacity	Cooling Efficiency	Heating Capacity	Heating Efficiency
R32/R125 (60/40 wt.%)	−3 to +7%	−10 to +5%	−3 to +4%	−3 to +1%
R32/R125 (50/50 wt.%)	−2 to +5%	+1 to +6%	n/a	n/a
R32/R125/R134a (30/10/60 wt.%)	−5 to +5%	−10 to +2%	−4 to +5%	−13 to 0%
R32/R125/R134a (23/25/52 wt.%)	−7 to +1%	−10 to −3%	−2 to +2%	−7 to +2%

Source: Godwin, 1993.

implement" substitute whereas R410A appears to be a "difficult-to-implement" substitute requiring major system redesign. R410A seems to be the leading long-term candidate for the new residential and light commerce equipment (Chin and Spatz, 1999).

By using the Mark V DOE/ORNL heat pump model, Chin and Spatz (1999) also noted that the performance of R410A is higher than R22 only when the ambient temperature is lower than 35°C due to the compressor efficiency degradation at the higher ambient temperature, as shown on Figures 7.2.2 and 7.2.3.

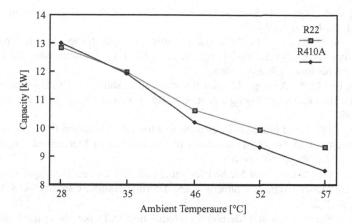

FIGURE 7.2.2 Ambient temperature vs. performance changes (Chin and Spatz, 1999).

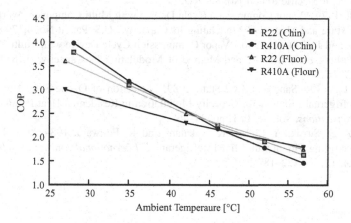

FIGURE 7.2.3 Ambient temperature vs. performance changes (Chin and Spatz, 1999).

REFERENCES

Chen, J. and H. Kruse, 1997, "Mass Fraction Shift Simulation for the Mixed Refrigerants R404A, R32/R134A, and R407C in an Air-conditioning System," *International Journal of HVAC&R Research*, Vol. 3, No. 2, pp. 149–157.

Chin, L. and M. W. Spatz, 1999, "Issues Relating to the Adoption of R410A in Air Conditioning Systems," Proceedings of 20th International Congress of Refrigeration, IIR/IIF, No. 179.

David, G. and M. Menzer, 1993, "Results of Compressor Calorimeter Tests in ARI's R22 Alternative Refrigerants Evaluation Program," Proceedings of ASHRAE/NIST Refrigerants Conference for R22/R502 Alternatives, pp. 1–18.

Fisher, S.K. and J.R. Sand, 1993, "Screening Analysis for Chlorine-Free Alternative Refrigerants to Replace R22 in Air-Conditioning Applications," *ASHRAE Transactions*, Vol. 99, Pt. 2, pp. 627–634.

Hughmark, G.A., 1962, "Holdup in Gas-Liquid Flow," *Chemical Engineering Progress*, Vol. 58, No. 4, pp. 62–65.

Jacobs, R. and H. Kruse, 1978, "The Use of Nonazeotropic Refrigerant Mixtures in Heat Pumps for Energy Saving," Proceedings of Meeting of Commission B2 of International Institute of Refrigeration.

Rothfleisch, P.I., 1995, "A Simple Method of Composition Shifting with a Distillation Column for a Heat Pump Employing a Zeotropic Refrigerant Mixture," NIST Report, NISTIR 5689.

Smith, S.L., 1969, "Void Fractions in Two-Phase Flow: A Correlation Based upon an Equal Velocity Head Model," Proceedings of the Institute of Mechanical Engineers 184, Pt. 1, No. 36, pp. 647–664.

Vakil, H.B., 1981, "Means and Method for Independently Controlling Vapor Compression Cycle Device Evaporator Superheat and Thermal Transfer Capacity," U.S. Patent No. 4,290,272.

Vakil, H.B., 1983, "Means and Method for Modulating and Controlling the Capacity of a Vapor Compression Cycle Device," U.S. Patent No. 4,384,460.

Vakil, H.B., 1983, "New Concept in Capacity Modulation using Nonazeotropic Mixtures," General Electric Report No. 83CRD212.

Vakil, H.B., 1979, "Vapor Compression Cycle Device with Multi Component Working Fluid Mixture and Method of Modulating Its Capacity," U.S. Patent No. 4,179,898.

Vakil, H.B. and J.W. Flock, 1980, "Vapor Compression Cycle Device with Multi Component Working Fluid Mixture and Method of Modulating Its Capacity," U.S. Patent No. 4,217,760.

Vineyard, E.A, J.R. Sand and T.G. Statt, 1989, "Selection of Ozone-Safe, Nonazeotropic Refrigerant Mixtures for Capacity Modulation in Residential Heat Pump," *ASHRAE Transactions*, Vol. 95, Pt.1, pp. 34–46.

Yoshida, Y., S. Suzuki, Y. Mukai, K. Nakatani and K. Fujiwara, 1989, "Development of Rectifying Circuit with Mixed Refrigerants," *International Journal of Refrigeration*, Vol. 12, pp. 182–187.

8 Heat Transfer of Refrigerant Mixtures

In this chapter, the heat transfer characteristics of refrigerant mixtures are discussed, including evaporation and condensation. The single-phase heat transfer of mixtures is affected only to the degree that the transport properties change. As can be expected from the phase equilibrium characteristics of mixtures, the heat transfer behavior of mixtures in the two-phase range is quite different from that of pure refrigerants. In this chapter, the various features of the heat transfer of mixtures are described and recent efforts of heat transfer measurements and the development of correlations are summarized.

8.1 NUCLEATE POOL BOILING HEAT TRANSFER COEFFICIENTS

The pool boiling behavior of zeotropic mixtures is somewhat different from pure refrigerants, as shown in Figure 8.1.1. Due to the difference in vapor and liquid mass fractions of the zeotropic mixture, the onset of boiling requires a higher wall superheating than that of a pure fluid. The critical heat flux for mixtures can be either higher or lower than that of pure fluids.

The measured pool boiling heat transfer coefficient of the zeotropic mixture is lower than that of the ideal pool boiling heat transfer coefficient predicted from that of pure fluids. Figure 8.1.2 shows the pool boiling heat transfer coefficient of R11/R113 mixtures on the outside surface of three different tubes at a pressure of 1 bar. The ideal heat transfer coefficient of zeotropic mixtures is defined here as the line connecting the two heat transfer coefficients of pure fluids, shown as a thick solid line in Figure 8.1.2 (Trewin et al., 1994). The heat transfer coefficient of zeotropic mixtures does not change linearly with mass fraction but it shows degradation compared to the ideal heat transfer coefficient. Moreover, the heat transfer coefficient of zeotropic mixtures shows much higher degradation in enhanced tubes. To understand this behavior of zeotropic mixtures, the processes that lead to the growth of a bubble must be considered.

Figure 8.1.3 explains the bubble growth mechanism in zeotropic mixtures. Since the mass fraction of the vapor inside the bubble is markedly different than that of the surrounding liquid — as discussed in Chapter 2 — the more volatile component must diffuse through the liquid to the growing bubble. The liquid layer immediately surrounding the bubble becomes depleted of the more volatile component and forms a diffusion area around the bubble. In order for the bubble to grow further, the more volatile component must diffuse through this layer of depleted liquid. The driving force is the superheating. The effective superheating of the liquid layer surrounding

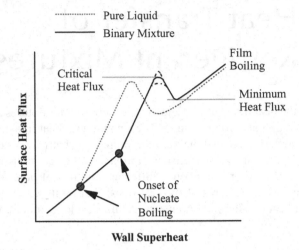

FIGURE 8.1.1 Schematic diagram of pool boiling curve for binary mixture.

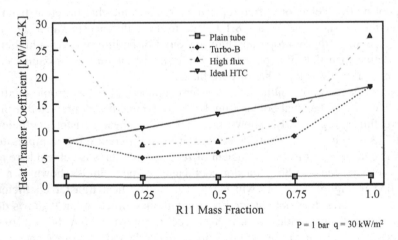

$P = 1$ bar $q = 30$ kW/m^2

FIGURE 8.1.2 Pool boiling heat transfer coefficients of R11/R113 mixtures (Trewin et al., 1994).

the bubble is no more than the temperature difference between the wall temperature and the bulk liquid saturation temperature for the given overall mass fraction. Since the liquid layer surrounding the bubble is enriched in the low vapor pressure component, its saturation temperature is higher than that of the bulk liquid. This reduction of the effective superheating is the main and most complex contribution to the heat transfer coefficient degradation.

The reasons for the heat transfer coefficient degradation of zeotropic mixtures is summarized as follows:

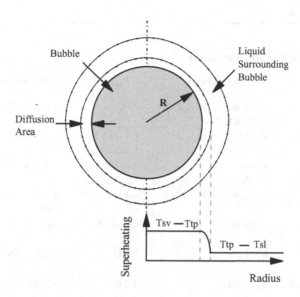

FIGURE 8.1.3 Bubble growth mechanism.

- Increase of the local saturation temperature because the more volatile component evaporates more readily than the other component during the bubble growth process as explained above (Van Wijk et al., 1956).
- Additional mass diffusion resistance of the more volatile component in the bubble (Sterning and Tichacek, 1961).
- Large changes of mixture physical properties with mass fraction (Stephan and Preusser, 1979).
- Lower boiling site densities caused by higher wall superheating (Grigor'ev, 1962).
- Retardation of the main heat transport mechanisms (Thome, 1983).
- Degradation of transport properties for mixtures as compared to pure components.

Since hydrofluorocarbons (HFCs) and their mixtures were introduced, research on nucleate pool boiling of these refrigerants has been conducted on plain tubes, low-finned tubes and enhanced tubes as summarized in Table 8.1 (Thome, 1996).

Ohata and Fujita (1994) showed the degradation of mixture heat transfer coefficients. Trewin et al. (1994) compared mixture heat transfer coefficients in three different tubes, as shown in Figure 8.1.2. Buschmeier et al. (1994), Schmidt (1995) and Gorenflo et al. (1995) investigated the heat transfer coefficients of the flammable refrigerant mixtures such as propane, butane and ethane.

8.2 FLOW BOILING HEAT TRANSFER COEFFICIENTS

Pool boiling relies on one major mechanism: bubbles are formed at a heated surface then detach and rise through an essentially stagnant pool to the surface. In flow

TABLE 8.1
Nucleate Pool Boiling Research

Authors	Refrigerant	P_{sat} or T_{sat}	Test Surface	Oil [%]
Webb and Pais (1991)	R123, R134a, R11, R12, R22, R123	27°C	PT, FT, TB, SE	None
Gorenflo and Sokol (1991)	R134a R227	1.4–36.5 bar 1.0–26.4 bar	PT, FT, TX	None
Shi and Hahne (1991)	R134a R152a	2.0–25.0 bar 2.2–23.0 bar	0.1 mm wire	None
Webb and Pais (1992)	R123, R134a, R11, R12, R22	4.4 and 26.7°C	PT, FT, TB, TX, SE	None
Gupte and Webb (1992)	R123, R11	4.4 and 26.7°C	SE	None
Memory et al. (1993)	R114	2.2°C	PT, FT, TB, HF	0–10
Webb and McQuade (1993)	R123, R11	4.4°C	PT, TB, SE	0–5
Muller et al. (1993)	R134a	4.4 and 32°C	0.1 mm wire 2.0 mm strip	0–10
Gupte and Webb (1994)	R123 R134a R11	4.4 and 26.7°C	TB, SE SE FT, TB, SE	None
Schomann et al. (1994)	Propane R114	0.7–17.1 bar 2.3–29.3 bar	PT	None
Buschmeier at al. (1994)	Propane/n-butane	3.8–21.1 bar	PT	None
Ohta and Fujita (1994)	R11/R113	1.0–15.0 bar	Flat plate	None
Trewin at al. (1994)	R11/R113	1.03 bar	PT, TB, HF	None
Schmidt (1995)	R22/R142b Propane/n-butane Ethane/n-butane	0.05–0.20[a] 0.03–0.90[a] 0.10–0.60[a]	Flat disk	None
Gorenflo et al. (1995)	Propane/i-butane/n-butane	1.6–12.6 bar	PT	None
Palm (1995)	R134a R22	0 and 20°C 0 and 25°C	PT, FT, TB, TX and Sevac	0–3

Note: PT: Plain tube, FT: Low-finned tube, TB: Turbo B tube, TX: Gewa TX 19 tube, SE: Gewa SE tube, HF: High flux,

[a] Reduced pressure.

Source: Thome, 1996.

boiling, the mechanism is more complex. Depending on flow conditions, there is still a significant boiling contribution but also a second one — evaporation. Whenever the tube is mostly filled with liquid or the film thickness of the liquid on the wall is high, the boiling contribution often dominates the heat transfer. As the vapor quality increases and the liquid film thickness is diminished, evaporation of molecules from the liquid surface becomes a more important contribution. When the liquid film becomes thinner than the diameter of a bubble before it detaches, the evaporation process dominates (suppression of nucleate boiling). Most flow boiling correlations try to tie these two contributions to the heat transfer process by having a boiling term and an evaporation term. The fact that the mass fraction of liquid and vapor phases are quite different for zeotropic mixtures affects these heat transfer processes quite significantly and must be accounted for in the correlations as well. Table 8.2 summarizes research on flow boiling heat transfer coefficients of mixtures based on Wang and Chato's (1995) summary, Thome's (1996) summary and an additional survey of the latest literature.

Ross et al. (1987) investigated convective boiling heat transfer coefficient of R152a/R13B1. Their results can be summarized as follows:

- Full suppression of nucleate boiling of pure fluids occurs only at low pressures.
- Mixtures can reach full suppression of nucleate boiling easier than pure fluids due to the mass transfer resistance.
- The heat transfer coefficients of mixtures are severely degraded (up to 50%) compared to pure fluids.
- For the evaporation dominated heat transfer regime, Chen's correlation applies well for both pure fluids and mixtures with a Prandtl number correction for pure fluids and without correction for mixtures. This means there is some nucleation for pure fluids while nucleate boiling is suppressed for mixtures.
- For the nucleate boiling dominated heat transfer regime, Stephan and Abdelsalam's (1981) correlation works well (within 20%) for pure fluids and Thome's (1983) correlation for mixtures.
- For the mixtures, there is a circumferential variation of the heat transfer coefficient opposite that found for pure fluids. This phenomenon means that there is a circumferential gradient in mass fraction and interfacial temperature.

Jung et al. (1989a,b) investigated the convective boiling heat transfer coefficient of R22/R114 and R12/R152a. Their results are summarized here:

- Flow boiling characteristics of zeotropic mixtures are different from those of azeotropic mixtures that behave similar to the pure fluids. Zeotropic mixtures have a mass diffusion effect due to the boiling temperature difference between the constituents.
- Flow boiling heat transfer coefficients of mixtures are lower than the ideal heat transfer coefficient based on the mixing rule for heat transfer coefficients of pure fluids for nucleate pool boiling (Figure 8.2.1). The actual mixture heat transfer coefficient at 65% quality was lower by 19, 36 and 35% than that of ideal values for the R22 compositions of 0.23, 0.47 and 0.77, respectively. Up to 80% of this heat transfer coefficient degradation is estimated to be due to the physical property variation. The mass transfer

TABLE 8.2
In-tube Flow Boiling Research

Authors	Refrigerant	Mass Flux [kg/m²-s]	Tube/ID [mm]
Singal et al. (1983)	R12/R13	n/a	Plain/9.5
Radermacher et al. (1983)	R13B1/R152a	n/a	Plain/9.1
Ross et al. (1987)	R13B1/R152a	n/a	Plain/9.1
Mulroy et al. (1988)	R22/R114, R13/R12	n/a	Annuli and ID Fins
Jung et al. (1989a, 1989b)	R22/R114, R12/R152a	250–720	Plain/9.1
Jung and Radermacher (1989)	R22/R114, R12/R152a	250–720	Plain/9.1
Hiraha et al. (1989)	R12/R22, R22/R114	100–350	Plain/8.0
Conklin and Vineyard (1990)	R143a/R124	150–350	Fluted/16.9
	R143a/R124	160–460	Finned/15.3
Murata and Hashizume (1990)	R11/R114	100–300	Plain/10.3
Eckels and Pate (1991b)	R22/R152a/R124	135–360	Plain/8.0
Sami and Duong (1991, 1992)	R22/R114	180–290	Fluted annuli/28.6 and 32.3
Sami and Schnotale (1992)	R22/R114, R22/R152a	142–255	Fluted/21.2
Sami et al. (1992)	R22/R152a/R114, R22/R152a/R124	142–255	Fluted/21.2
Niederkruger et al. (1992)	R846/R12	80–400	Plain/14.0
Doerr and Pate (1993)	R32/R134a	100–400	Plain/8.0
	R32/R125/R134a	125–360	
Torikoshi and Ebisu (1993)	R32/R134a	100–400	Plain/8.7
Torikoshi and Ebisu (1993)	R32/R125	100–400	Plain/N/a
Wattelet et al. (1993)	R32/R125	n/a	Plain/7.7
Murata and Hashizume (1993)	R123/R134a	100–300	Plain/10.3 Microfin/10.7
Wattelet et al. (1994a)	R22/R124/R152a	25–100	Plain/7.04
Kattan at al. (1994)	R502	100–300	Plain/12.0
	R125/R290/R22	102–318	
	R125/R143a/R134a	102–320	
Rohlin (1994)	R22/R142b, R32/R134a	30–160	Plain/15.0
Bivens and Yokozeki (1994)	R32/R134a, R32/R125/R134a	200	Plain/8.0
Melin and Vamling (1995)	R22/R142b, Propane/n-butane	85–316	Plain/10.0
Goto et al. (1995)	R32/R125	80–433	Microfin/6.4 and 9.5
	R32/R125/R134a	100–365	
Wijaya and Spatz (1995)	R32/R125	160–561	Plain/7.8
Uchida et al. (1996)	R32/R125/R134a	100–500	Plain, Microfin/6.4
Sundaresan et al. (1996)	R32/R125, R32/R125/R134a	125–375	Plain, Microfin/9.5
Shin et al. (1997)	R32/R134a, R32/125	424–741	Plain/7.7
	R290/R600a	265–583	

Note: ID: Inside diameter, OD: Outside diameter.

Sources: Wang and Chato, 1995; Thome, 1996.

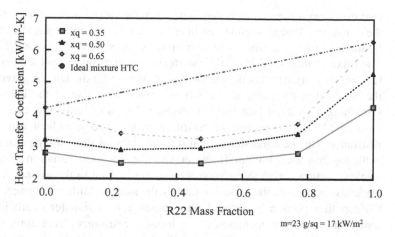

FIGURE 8.2.1 In-tube boiling heat transfer coefficients of R22/R114 mixtures (Jung et al., 1989a,b).

resistance is estimated to contribute less than 20% of the heat transfer coefficient degradation.

- Flow boiling heat transfer coefficients of mixtures are a function of the mixture mass fraction, vapor quality and heat flux (Figure 8.2.2). A steep gradient near pure R22 is due to the physical property variation. When the vapor quality is less than 20 to 30%, the nucleate boiling contribution is dominant; therefore, the heat transfer coefficient increases as the heat flux increases in this partial boiling regime. In the partial boiling regime, both the nucleate and convective boiling contribute. Moreover, the heat transfer coefficient decreases due to the reduction of the nucleate boiling contribution as the vapor quality increases.

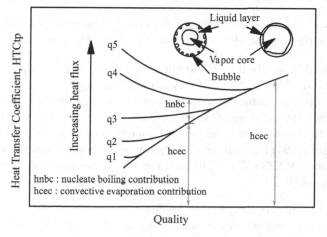

FIGURE 8.2.2 Boiling heat transfer mechanisms (Jung et al., 1989a,b).

- However, convective boiling is dominant when the vapor quality is high. For the convective boiling regime, the heat transfer coefficient is not affected by the heat flux since the nucleate boiling is suppressed (Figure 8.2.2).
- The mixture shows a different circumferential variation in wall temperature and heat transfer coefficient from that of pure fluids. Due to gravity, the liquid film thickness at the top is thinner than that at the bottom (assuming a constant heat flux). Therefore, for pure fluids, the top part has better heat transfer than the bottom part due to reduced heat transfer resistance. This heat transfer variation results in a lower wall temperature at the top than at the bottom. However, this film thickness variation causes a mass fraction variation for zeotropic mixtures. Due to the thinner film thickness at the top, the mass fraction of the more volatile component in the top film portion is reduced and causes a heat transfer coefficient reduction due to an increased mass transfer resistance. Therefore, the bottom portion of the liquid film provides better heat transfer than the top portion for zeotropic mixtures. This heat transfer variation results in a lower wall temperature at the bottom than at the top.

Shin et al. (1997) investigated convective boiling heat transfer coefficient of R32/R134a (25/75 wt.%), R290/R600a and R32/R125 (50/50 wt.%). Their results are:

- Similar to Jung et al.'s (1989) results, the heat transfer coefficient increases as the heat flux increases in this partial boiling regime and the heat transfer coefficient is not affected by the heat flux for the convective boiling regime.
- The heat transfer coefficient increases when the quality is higher than 20% as the refrigerant mass flux increases due to the increased mean velocity of the refrigerant.
- Similar to Jung et al.'s (1989) results, the local heat transfer coefficient of the zeotropic mixture at the top is lower than that at the bottom. However, the local heat transfer coefficient of the azeotropic mixture at the top is higher than that at the bottom as is the case for pure fluids.

Many other investigators confirmed the boiling heat transfer coefficient degradation as compared to the ideal heat transfer coefficient as follows:

- Torikoshi and Ebisu (1993) studied the boiling heat transfer coefficient of R32/R134a (30/70 wt.%) and reported the degradation of the boiling heat transfer coefficient as compared to the ideal boiling heat transfer coefficient was 30 and 20%, respectively, for the mass flux of 91 kg/m^2-s and 182 kg/m^2-s when the saturation temperature was 0°C.
- Rohlin (1994) reported the degradation of the boiling heat transfer coefficient of R22/R142b (60/40 wt.%) as 40% when the saturation temperature was 0°C.

Some researchers investigated the boiling heat transfer coefficient of binary and ternary mixtures that are considered as replacements of existing pure refrigerants, R12 and R22:

- Sami et al. (1992) studied the boiling heat transfer coefficient of R22/R152a/R114 (53/19/28 wt.%) and R22/R152a/R124 (50/20/30 wt.%) in an enhanced tube and reported the boiling heat transfer coefficient of these two ternary mixtures is lower than R22 and higher than R12. The boiling heat transfer coefficient of R22/R152a/R124 is higher than that of R22/R152a/R114 (53/19/28 wt.%) because the heat transfer coefficient of R124 is better than that of R114.
- Wattelet et al. (1994a) studied the boiling heat transfer coefficient of R22/R124/R152a (52/33/15 wt.%) and compared it with that of R12 and R134a. For an annular flow (mass flux: 200–500 kg/m²-s, heat flux: 5–30 kW/m²), the mixture has a heat transfer coefficient similar to R134a and 10% higher heat transfer coefficient than that of R12. When the heat flux is low, its nucleate boiling is more suppressed than R134a or R12. For a stratified flow (mass flux: 25–100 kg/m²-s, heat flux: 2–10 kW/m²), the mixture has a heat transfer coefficient similar to R134a. As the mass flux decreases, the heat transfer coefficients of the mixture are much lower than R12.
- Wijaya and Spatz (1995) reported that the boiling heat transfer coefficient of R410A is much higher (42 to 63%) than that of R22.
- Uchida et al. (1996) studied the boiling heat transfer coefficient of R32/R125/R134a (30/10/60 wt.%) and R22 inside horizontal tubes with smooth, grooved and cross-grooved inner surfaces. They found that the cross-grooved tube has the highest heat transfer coefficient with a mixture nearly three times higher than that of the smooth tube and 20 to 40% higher than that of the grooved tube. However, the heat transfer coefficient of R22 in the cross-grooved tube is lower than that of the grooved tube. They attributed this difference to the stirring effect of the liquid phase by the notches that reduce the composition variation in the liquid phase of the mixture.
- Sundaresan et al. (1996) reported their comparisons of the boiling heat transfer coefficients of R22, R410A and R407C in a smooth tube and microfin tube. They found that the heat transfer coefficients of R410A are the highest. The heat transfer coefficients of R410A is 25% higher than R22 and 72% higher than R407C in the smooth tube; the heat transfer coefficients of R410A is 29% higher than R22 and 100% higher than R407C in the smooth tube.

8.3 CORRELATIONS FOR FLOW BOILING HEAT TRANSFER

For modeling purposes, accurate heat transfer correlations for mixtures are needed to properly design heat exchangers. This section presents literature survey results of flow boiling heat transfer correlations.

A considerable number of heat transfer correlations of flow boiling for pure fluids have been proposed. However, only a few correlations are published for mixtures. The applicability of these general correlations for mixtures must be proven. Two review papers by Webb and Gupte (1992) and Darabi et al. (1995) offered a summary of the most referenced correlations for flow boiling. Webb and Gupte classified heat transfer models into three categories: superposition, enhancement and asymptotic models. In this section, twelve nondimensional general correlations are examined.

Chen, Bennett-Chen, Gungor-Winterton and Jung's correlations utilize a super-position model. In this model, the total heat transfer coefficient is the sum of nucleate boiling and bulk convection contributions. Shah, Schrock-Grossman, Mishra, Sami and Shin's correlations utilize an enhancement model. In this model, the two-phase heat transfer coefficient enhances the single-phase heat transfer coefficient of a flowing liquid by a two-phase enhancement factor. Liu-Winterton, Wattelet and Bivens-Yokozeki's correlations utilize an asymptotic model. In this model, the total heat transfer coefficient is of the form of Kutateladze's (1961) power-type addition.

8.3.1 SUPERPOSITION MODELS

8.3.1.1 Chen Correlation

Chen (1966) proposed a general correlation for saturated boiling in a vertical tube. His correlation is based on the superposition model, which divides the heat transfer into two parts: a nucleate pool boiling contribution (h_{nb}) based on Foster and Zuber's pool boiling equation (1955) and a bulk convective contribution (h_{bc}) based on Dittus and Boelter's (1930) equation, as shown in Equations 8.1 through 8.3,

$$h = h_{nb} + h_{bc} \tag{8.1}$$

$$h_{bc} = 0.023 \frac{k_l}{D} \mathrm{Re}_{tp}^{0.8} \mathrm{Pr}_l^{0.4} \tag{8.2}$$

where
Re_{tp} = two-phase Reynolds number

$$\mathrm{Re}_{tp} = \frac{G(1-x)D}{\mu_l} F(X_{tt})^{1.25}$$

where
$F(X_{tt})$ = two-phase correction factor

$F(X_{tt}) = 1.0$ for $X_{tt} \geq 10$
$F(X_{tt}) = 2.35(0.213 + X_{tt}^{-1})^{0.736}$ for $X_{tt} < 10$
$\quad X_{tt}$ = Martinelli's parameter

$$X_{tt} = \left\{ (1-x)/x \right\}^{0.9} \left(\frac{D_v}{D_l} \right)^{0.5} \left(\frac{\mu_v}{\mu_l} \right)^{0.1}$$

$$h_{nb} = 0.00122 \left(\frac{k_l^{0.79} C p_l^{0.45} \rho_l^{0.49}}{\sigma^{0.5} \mu_l^{0.29} h_{fg}^{0.24} \rho_{fg}^{0.24}} \right) \left[T_w - T_{sat}(P_l) \right]^{0.24} \left[P_{sat}(T_w) - P_l \right]^{0.75} S \quad (8.3)$$

where
  S = two-phase correction factor

$$S(\mathrm{Re}_{tp}) = \left(1 + 2.56 \times 10^{-6} \mathrm{Re}_{tp}^{1.17} \right)^{-1}$$

In this correlation, the nucleate boiling term is suppressed by a two-phase correction factor (S) as the vapor quality increases. This factor accounts for the thinner boundary layer at a higher vapor quality. The bulk convection term is enhanced by a two-phase correction factor $F(X_{tt})$ as the vapor quality increases. This factor accounts for the higher vapor velocities at the higher vapor quality. The Chen correlation was developed to fit flow boiling data for water, methanol, cyclohexane and pentane.

8.3.1.2 Bennett and Chen Correlation

Bennett and Chen (1980) modified the Chen correlation to account for the analysis of nucleate boiling near the wall and for the effects of a liquid Prandtl number higher than 1 on the bulk convection. Moreover, this correlation is based on experiments with an aqueous mixture of ethylene glycol and water in a vertical channel. The bulk convection term (h_{bc}) was modified, as shown in Equation 8.4. The two-phase correction factor was also modified, as shown in Equation 8.5,

$$h_{bc} = h_l F(X_{tt}) \mathrm{Pr}_l^{0.296} \quad (8.4)$$

where
  h_l = liquid phase heat transfer coefficient

$$h_l = 0.023 \frac{k_l}{D} \mathrm{Re}_l^{0.8} \mathrm{Pr}_l^{0.4}$$

$$S = \frac{1 - \exp\left(\frac{-F(X_{tt}) h_l X_o}{k_l} \right)}{\frac{F(X_{tt}) h_l X_o}{k_l}} \quad (8.5)$$

where $X_o = 0.04 \left(\frac{\sigma}{g(\rho_l - \rho_v)} \right)^{0.5}$

As shown in Equation 8.5, this correlation includes the surface tension and density difference between liquid and vapor phases in the nucleate boiling suppression factor. The heat transfer enhancement factor (F) includes thermal characteristics of the fluid.

Ross et al. (1987) suggested the use of Bennett and Chen correlation with the two-phase correction factor S set to zero. They proposed this idea based on their experiment on the convective boiling heat transfer coefficient of R152a/R13B1. Since the nucleate boiling is suppressed, the evaporation is the main heat transfer mechanism for the binary mixtures.

8.3.1.3 Gungor and Winterton Correlation

Gungor and Winterton (1986) developed the following correlation for flow boiling in horizontal and vertical tubes. They used the basic form of the Chen correlation in superposition form, as shown in Equation (8.6):

$$h = Eh_l + Sh_{pool} \qquad (8.6)$$

where

$\qquad h_l$ = heat transfer coefficient for the liquid phase flowing alone
$\qquad h_{pool}$ = pool boiling heat transfer coefficient
$\qquad E$ = forced convection enhancement parameter
$\qquad S$ = forced convection suppression parameter

$$h_{pool} = 55 P\gamma^{0.12}(-\log_{10} P\gamma)^{-0.55} M^{-0.5} q^{0.67}$$

$$E = 1 + 24000 Bo^{1.16} + 1.37 X_{tt}^{-0.86}$$

$$S = (1 + 1.15 \times 10^{-6} E^2 \operatorname{Re}_l^{1.17})^{-1}$$

In this correlation, two-phase correction factors (E and S) were obtained iteratively from over 4300 data points for water, various refrigerants and ethylene glycol, covering seven fluids. Gungor and Winterton chose the Cooper (1984) correlation for the pool boiling heat transfer coefficient.

8.3.1.4 Jung Correlation

Jung et al. (1989a) developed the following flow boiling correlation for horizontal tubes for the binary mixtures R22/R114 and R12/R152a. They used the basic form of the Chen correlation in superposition form similar to Gungor and Winterton's as shown in Equation 8.7,

$$h = C_{me} F h_l + S/C_{un} h_{pool} \qquad (8.7)$$

where

$\qquad C_{me} = 1 - 0.35|X_v - X_l|^{1.56}$
$\qquad F$ = forced convection enhancement parameter

$F = 2.37(0.29 + 1/X_{tt})^{0.85}$

h_l = heat transfer coefficient for the liquid phase flowing alone (Equation 8.5)

S = forced convection suppression parameter

$S = 4048X_{tt}^{1.22}B_O^{1.13}$ for $X_{tt} < 1$

$S = 2 - 0.1X_{tt}^{-0.28}B_O^{-0.33}$ for $1 < X_{tt} < 5$

$C_{un} = |1 + (b_2 + b_3)(1 + b_4)|(1 + b_5)$

h_{pool} = pool boiling heat transfer coefficient

$h_{pool} = h_i/C_{un}$

where

$$h_i = h_a h_b/(h_b X_l + h_a X_v)$$

X_l = liquid phase composition

X_v = vapor phase composition

$h_a h_b$ = pool boiling heat transfer coefficient of pure refrigerants a and b given
by Stephan and Abdelsalam (1981) correlation

$b_2 = (1 - X_l) \ln|(1.01 - X_l)/(1.01 - X_v)| + X_l \ln(X_l/X_v) + |X_l - X_v|^{1.5}$

$b_3 = 0$ for $X_l \geq 0.01$

$b_3 = (X_l/X_v)^{0.1} - 1$ for $X_l < 0.01$

$b_4 = 152(P/P_{ca})^{3.9}$

$b_5 = 0.92|X_l - X_v|^{0.001} + (P/P_{ca})^{0.66}$

$X_l/X_v = 1$ for $X_l = X_v = 0$

where

P_{ca} = critical pressure of more volatile component

Jung et al. reported that this correlation matches their experimental data for R22/R114 (23/77, 48/52, 77/23 wt.%) over 1700 data points and R12/R152a (21/79, 88/12 wt.%) within 9.6% deviation.

8.3.2 ENHANCEMENT MODELS

8.3.2.1 Shah Correlation

For saturated flow boiling in vertical and horizontal tubes, Shah (1976) proposed the enhancement model, which enhances the single-phase heat transfer coefficient of flowing liquid by a two-phase enhancement factor as shown in Equation 8.8. He classified the flow-boiling regime into three regions by utilizing three dimensionless parameters, shown in Equation 8.8. They are the pure nucleate boiling regime, the bubble suppression regime and the convective boiling regime:

$$\psi_s = h/h_l = f\left(Co, Bo, Fr_{le}\right) \qquad (8.8)$$

where
$$Co = [(1 - x)/x]^{0.8} [\rho_v/\rho_l]^{0.5}$$
$$Bo = (q^{11})/(Gh_{lv})$$
$$Fr_{le} = (G^2)/(\rho_l^2 gD)$$

Initially, he presented this correlation graphically; later, he recommended the following computational representation of his correlation as shown in Equation 8.9 (Shah, 1982):

$$Ns = Co \text{ for } Fr_{le} \geq 0.04 \tag{8.9}$$

$$Ns = 0.38Fr_{le}^{-0.3} Co \text{ for } Fr_{le} < 0.04$$
$$Fs = 14.7 \text{ for } Bo \geq 11 \times 10^{-4}$$
$$Fs = 15.4 \text{ for } Bo < 11 \times 10^{-4}$$
$$y_{cb} = 1.8Ns^{-0.8} \text{ for } Ns > 1.0$$
$$y_{nb} = 230Bo^{0.5} \text{ for } Bo > 0.3 \times 10^{-4}$$
$$y_{nb} = 1 + 46Bo^{0.5} \text{ for } Bo \leq 0.3 \times 10^{-4} \text{ for } Ns \leq 1.0$$
$$y_{bs} = Fs\, Bo^{0.5} \exp(2.74\, Ns^{-0.1}) \text{ for } 0.1 < Ns \leq 1.0$$
$$y_{bs} = Fs\, Bo^{0.5} \exp(2.74\, Ns^{-0.15}) \text{ for } Ns \leq 0.1$$
$$y_s = \text{the larger of } y_{nb} \text{ and } y_{cb} \text{ for } Ns > 1.0$$
$$y_s = \text{the larger of } y_{bs} \text{ and } y_{cb} \text{ for } Ns \leq 1.0$$

Wattelet et al. (1994a) examined Shah's correlation with their experimental data for R22/R124/R152a and reported that Shah's correlation overestimates the experimental data by more than 50% in the wavy-stratified flow regime, but matches well with the experimental data in the annual flow regime.

8.3.2.2 Schrock and Grossman Correlation

Schrock and Grossman (1959) developed the following correlation based on a fit to water in a vertical tube. This correlation is an enhancement model like Shah's as shown in Equation 8.10,

$$h/h_l = C_1\left(Bo + C_2 X_{tt}^{-0.66}\right) \tag{8.10}$$

where
Bo = boiling number
$C_1 = 7.39 \times 10^{-3}$
$C_2 = 1.5 \times 10^{-4}$

8.3.2.3 Mishra Correlation

Mishra et al. (1981) developed the following correlation based on their data fit to R12/R22. This correlation is an enhancement model like Shah's and very similar to Schrock and Grossman's, as shown in Equation 8.11,

$$h/h_l = C_1 Bo^n X_{tt}^{-m} \qquad (8.11)$$

where

C_l = 5.64 for 23–27 wt.% of R12 and 21.75 for 41–48 wt.% of R12
m = 0.23 for 23–27 wt.% of R12 and 0.29 for 41–48 wt.% of R12
n = 0.15 for 23–27 wt.% of R12 and 0.23 for 41–48 wt.% of R12

8.3.2.4 Sami Correlation

Sami et al. (1992) developed their correlation Equation 8.12 for mixtures in an enhanced tube — that is, an enhancement model in a form of the Dittus and Boelter equation. However, Sami's correlation uses a thermal conductivity term instead of the Prandtl number.

$$h = 0.155(k_l/D) A \, Re^{0.62} k_l^{0.3} \qquad (8.12)$$

where

k_l = thermal conductivity of liquid
$A = 1 - C(\Sigma|X_L - X_V|)^{0.82}$
$C = 0.79$ for mixtures with HFCs
$C = 0.68$ for mixtures with HCFCs

Sami et al. (1992) developed this correlation with R22/R152a/R114 (53/19/28 wt.%) and R22/R152a/R124 (50/20/30 wt.%). Later, Sami et al. (1995) reported that this correlation matches their experimental data for R23/R22/R134a (5/90/5 wt.%), R23/R22/R152a (5/90/5 wt.%), R23/R125/R152a (8/62/30 wt.%) and R23/R22/R134a (5/65/30 wt.%) within a 20% deviation.

8.3.2.5 Shin Correlation

Shin et al. (1996) developed the following correlation Equation 8.1 based on a fit to R32/R134a and R290/R600a data of three compositions (75/25, 50/50, 25/75 wt.%). This correlation is an enhancement model like Shah's and used a correction factor to account for the mass transfer effect due to the composition difference between liquid and vapor phases, considered a driving force for mass transfer at the interface:

$$h/h_l = (1 - C_F)F \qquad (8.13)$$

where

F = Chen's two-phase correction factor
$C_F = A|Y - X|^n$

where

A = curved fitted coefficients (0.569 for R32/R134a; 0.533 for R290/R600a)
n = curved fitted coefficients (0.860 for R32/R134a; 0.828 for R290/R600a)

     X = liquid phase composition
     Y = vapor phase composition

This correlation matches with experimental data within a 9% deviation for R32/R134a and 15% for R290/R600a.

8.3.3 Asymptotic Models

8.3.3.1 Liu and Winterton Correlation

Liu and Winterton (1991) proposed the following correlation for flow boiling inside tube and annuli. They utilized an asymptotic model based on Kutateladze's (1961) power-type addition, as shown in Equation 8.14,

$$h = \sqrt{(Eh_l)^2 + (Sh_{pool})^2} \tag{8.14}$$

where

     $E = [1 + x\mathrm{Pr}_l \, (\rho_l/\rho_v - 1)]^{0.35}$
     $S = (1 + 0.055E^{0.1} \, \mathrm{Re}_l^{0.16})^{-1}$

In this correlation, two-phase correction factors (E and S) were obtained from 4300 data points for water, various refrigerants and ethylene glycol fluids. The Cooper (1984) correlation is used as a pool boiling heat transfer coefficient.

8.3.3.2 Wattelet Correlation

Wattelet et al. (1994b) proposed an asymptotic model based on their data for R12 and R134a. The Wattelet correlation accounted for the decrease in heat transfer due to the wavy-stratified flow pattern in the low mass fluxes by adding a Froude dependence to the convective term, as shown in Equation 8.15,

$$h = \left[h_{nb}^{\,2.5} + h_{bc}^{\,2.5} \right]^{1/2.5} \tag{8.15}$$

where

$$h_{nb} = 55M^{-0.5}q^{0.67}P_r^{0.12}[-\log_{10} P_r]^{-0.55}$$

where

     M = molecular weight
     q = heat flux [W/m^2]
• Pr = reduced pressure

$$h_{bc} = Fh_l \, R$$

where

$$F = 1 + 1.925X_{tt}^{-0.83}$$
$$h_l = 0.23\mathrm{Re}_x^{0.8}\,\mathrm{Pr}^{0.4}\,\lambda_l/D$$
$$R = 1.32\mathrm{Fr}_l^{0.2} \text{ if } \mathrm{Fr}_l < 0.25$$
$$R = 1 \text{ if } \mathrm{Fr}_l \geq 0.25$$

where

$$\mathrm{Fr}_l = (G/\rho_l)^2/(9.80665D)$$

8.3.3.3 Bivens and Yokozeki Correlation

Bivens and Yokozeki (1994) suggested a modified Jung correlation and Wattelet correlation for R32/R125 and R32/R125/R134a to account for the mass transfer resistance, as shown in Equation 8.16,

$$h = h_{tp}/\left\{1 + h_{tp}T_{int}/q\right\} \qquad (8.16)$$

where

q = heat flux [W/m²]
T_{int} = temperature at the liquid-vapor interface

$$T_{int} = 0.175(T_d - T_b)\left\{1 - \exp\left(-\frac{q}{1.3\times10^{-4}\rho_l h_{fg}}\right)\right\}$$

where

T_d = dew point temperature [°C]
T_b = bubble point temperature [°C]
ρ_l = liquid density [kg/m³]
h_{fg} = heat of vaporization [J/kg]
h_{tp} = heat transfer coefficient

$$h_{tp} = \left[h_{nb}^{2.5} + h_{bc}^{2.5}\right]^{1/2.5}$$

where

$$h_{nb} = 55M^{-0.5}q^{0.67}P_r^{0.12}[-\log_{10}P_r]^{-0.55}$$

$$h_{bc} = Fh_l R$$

where

$$F = (0.29 + 1/X_{tt})^{0.85}$$
$$R = 2.838\mathrm{Fr}_l^{0.2} \text{ if } \mathrm{Fr}_l \leq 0.25$$
$$R = 2.15 \text{ if } \mathrm{Fr}_l > 0.25$$

8.3.4 COMPARISON OF CORRELATIONS

Judge (1996) compared five flow boiling correlations in his cycle simulation study to determine which works best with the simulation to predict the overall heat transferred for residential air-conditioning and heat pumping applications with R22 and R407C. The flow boiling heat transfer correlations examined in his study were Chen (1966), Gungor and Winterton (1986), Shah (1982), Jung (1989a) and Kandlikar (1991). Jung and Kandlikar's correlations were the only ones among these five correlations developed for mixtures. These correlations are generally applicable to any binary mixture of halogenated refrigerants.

The simulation results are presented by plotting the measured capacity vs. the capacity predicted by the model. The evaporation results are shown in Figure 8.3.1. When using Jung's heat transfer correlation (providing the best results), the maximum difference between the simulation and experimental results ranges from −5.0 to +3.8%. Figures 8.3.2 and 8.3.3 show the flow boiling heat transfer coefficient vs. position for the different heat transfer correlations for R22 and R407C, respectively. Note that in Figure 8.3.3 all but one of the pure component heat transfer correlations predict lower heat transfer coefficients than does Jung. This was not anticipated because the pure component heat transfer correlations do not take into account the degradation in heat transfer associated with mixtures. The other correlation developed for mixtures, Kandlikar's, predicts heat transfer coefficients that are significantly lower than that of Jung. This is particularly interesting since Kandlikar used Jung's data to develop his correlation.

8.4 PRESSURE DROP DURING EVAPORATION

In this section, some of the recent experimental data and correlation research on horizontal boiling flow and pressure drop of mixtures are summarized. The pressure drop in a pipe consists of three major contributions: frictional pressure drop (ΔP_f), accelerational pressure drop (ΔP_a) and gravitational pressure drop (ΔP_g). For the horizontal tube, the gravitational pressure drop is eliminated ($\Delta P_g = 0$). Moreover, the accelerational pressure drop is generally much smaller than the frictional pressure drop; therefore, most investigators only consider the frictional pressure drop. Two major parameters of the frictional pressure drop are the liquid viscosity and mass flux.

8.4.1 MARTINELLI AND NELSON CORRELATION

The Martinelli and Nelson (1948) correlation is one of the best accepted correlations for the pressure drop during convective boiling in a horizontal tube, shown in Equation 8.17. This correlation is based on a two-phase slip flow model:

$$\Delta P_{MN} = \frac{2f_{f0}G^2L}{D\rho_l}\left[\frac{1}{x}\int_0^x \phi_{f0}^2 dx\right] + \frac{G^2}{\rho_l}\left[\frac{x^2}{\alpha}\left(\frac{\rho_l}{\rho_v}\right) + \frac{(1-x)^2}{(1-\alpha)} - 1\right] \qquad (8.17)$$

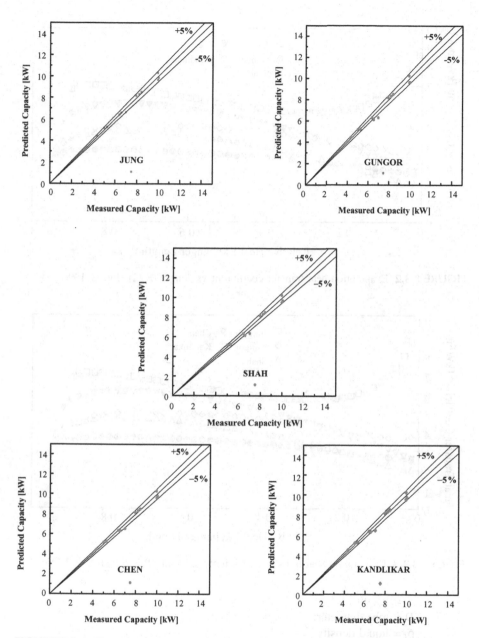

FIGURE 8.3.1 Flow boiling heat transfer correlation results (Judge, 1996).

where

ΔP_{MN} = Martinelli and Nelson total two-phase pressure drop

f_{f0} = frictional coefficient considering two-phase flow as liquid flow

G = mass flux

x = vapor quality

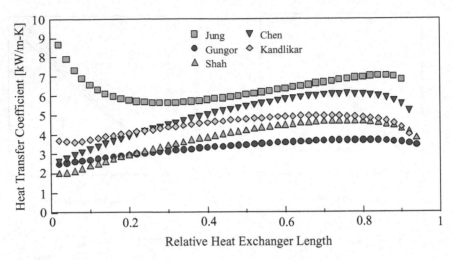

FIGURE 8.3.2 Evaporation heat transfer coefficient vs. length (R22) (Judge, 1996).

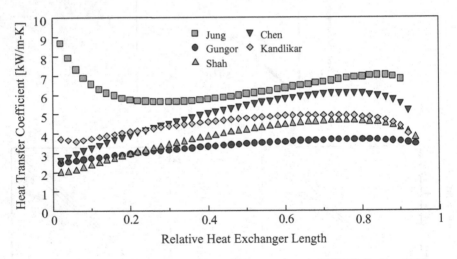

FIGURE 8.3.3 Evaporation heat transfer coefficient vs. length (R407C) (Judge, 1996).

L = length
D = tube diameter
ρ_l = liquid density
ρ_v = vapor density
ϕ_{f0}^2 = two-phase frictional multiplier
α = void fraction

The Blasius friction factor relation for the turbulent flow in a smooth pipe as shown in Equation 8.18 can be used to obtain the frictional coefficient (f_{f0}):

$$f_{fo} = 0.079 \, \text{Re}^{-0.25} \tag{8.18}$$

The frictional pressure drop when the total flow is liquid is given in Equation 8.19,

$$\Delta P_{f,MN} = \frac{2f_{fo}G^2 L}{D\rho_l} \phi_{fo}^2 \tag{8.19}$$

To obtain the total pressure drop during the evaporation, the Martinelli and Nelson correlation requires the evaluation of the local two-phase frictional multiplier (ϕ_{2fo}) and void fraction (α). Wallis (1969) gave the following relations (Equations 8.20 and 8.21) for these two parameters:

$$\phi_{fo}^2 = \phi_l^2 (1-x)^{1.75} \tag{8.20}$$

where $\phi_1 = \left(1 + \frac{1}{x^{0.5}}\right)^2$

$$\alpha = (1 + x^{0.8})^{-0.378} \tag{8.21}$$

Ross et al. (1987) compared their experimental data for R13B1/R152a (18/82, 38/62 and 80/20 wt.%) with Chisholm's (1967) correlation, a modified version of the Martinelli and Nelson method. They reported that the Chisholm correlation matches with their data within a 30% variation and there is no need to modify the Chisholm correlation.

8.4.2 SINGAL CORRELATION

Singal et al. (1983) studied the pressure drop of R12 and R13/R12 (5/95, 10/90 and 15/85 wt.%) during evaporation and reported that the Martinelli and Nelson correlation overpredicts the pressure drop of R12 and underpredicts the pressure drop of R13/R12 mixtures. Singal et al. then suggested two sets of total pressure drop (ΔP_{tp}) correlations (Equations 8.22) and two sets of frictional pressure drop (ΔP_f) correlations (Equations 8.23) by modifying the Martinelli and Nelson correlation:

$$\Delta P_{tp} = 0.87(1+C)^{2.66} \Delta P_{MN}$$
$$\Delta P_{tp} = 0.89(1-C)^{-2.12} \Delta P_{MN} \tag{8.22}$$

$$\Delta P_f = 0.86(1+C)^{2.86} \Delta P_{f,MN}$$
$$\Delta P_f = 0.88(1-C)^{-2.28} \Delta P_{f,MN} \tag{8.23}$$

Singal et al. reported that these correlations match their data within a 30% deviation. However, these correlations cannot be used for R13 mass fractions larger than 25%.

8.4.3 SAMI CORRELATION

Sami et al. (1992) suggested the following correlation (Equation 8.24) using the Bo Pierre (1954) form based on their measurement of R22/R152a/R114 and R22/R152a/R124 in an enhanced tube. The Pierre correlation is a semi-empirical correlation based on a two-phase homogeneous model. Sami's correlation matches the experimental data within a 15% deviation. Later, Sami et al. (1995) examined their correlation against their data for additional mixtures R23/R22/R152a (5/90/5, 5/65/30 wt.%), R23/R22/R290 (5/90/5 wt.%), R23/R22/R134a (5/65/30 wt.%) and R23/R125/R152a (8/62/30 wt.%) and reported the mean deviation is 30% for most of the data.

$$\Delta P_{tp} = \left\{ f_{avg} + \frac{(x_2 - x_1)D}{x_{avg}L} \right\} \frac{G^2 L}{D\rho_{avg}} \tag{8.24}$$

where

$\quad f_{avg} = 0.43\, K_f^{\,0.3}\, \mathrm{Re}^{-0.28}$
$\quad K_f = $ Bo Pierre number
$\quad K_f = \Delta x\, h_{fg}/L$
$\quad \rho_{avg} = \rho_v\, \rho_l\, [x_{avg}\, \rho_l + (1 - x_{av})\, \rho_v]$
$\quad x_{avg} = (x_i + x_o)/2$

8.4.4 JUNG AND RADERMACHER CORRELATION

Jung and Radermacher (1989) studied the pressure drop of R12, R22, R114, R152a and R12/R152a. They reported that the Pierre correlation is not applicable to half of their data and the Martinelli and Nelson correlation overpredicted their data by 20% but is applicable. The authors suggested a modified version of the Martinelli and Nelson correlation, as shown in Equation 8.25. Their correlation matches their data for both pure fluids and mixtures within an 8% deviation.

$$\Delta P_{tp} = \frac{2 f_{f0} G^2 L}{D\rho_l} \left[\frac{1}{\Delta x} \int_{x1}^{x2} \phi_{tp}^2 dx \right] \tag{8.25}$$

where

$\quad f_{fo} = 0.079\, \mathrm{Re}^{-0.25}$

$\quad \phi_{tp}^2 = 30.78 x^{1.323} (1 - x)^{0.477} P_r^{-0.7232}$

8.4.5 Souza and Pimenta Correlation

Souza and Pimenta (1995) studied the pressure drop of R12, R22, R134a, R152a, R22/R152a/R124 and R32/R125. They reported that flow regimes were predominantly annular flow and stratified-wavy flow for high mass fluxes and low mass fluxes, respectively. The authors suggested a modified version of Chrisholm's (1973) separate flow model for a frictional pressure drop, as shown in Equation 8.26. Their correlation matches their data for both pure fluids and mixtures within an 8% deviation.

$$\Delta P_{tp} = \frac{2f_{f0}G^2L}{D\rho_l}\left[\frac{1}{\Delta x}\int_{x1}^{x2}\phi_{tp}^2 dx\right]$$ (8.26)

where

$$f_{fo} = 0.079\,\mathrm{Re}^{-0.25}$$

$$\phi_{tp}^2 = 1 + (\Gamma^2 - 1)x^{1.75}(1 + 0.9524\Gamma X_{tt}^{0.4126})$$

where

$$\Gamma = \left(\frac{\rho_l}{\rho_v}\right)^{0.5}\left(\frac{\mu_v}{\mu_l}\right)^{0.125}$$

8.4.6 Other Studies

There are two interesting issues regarding the pressure drop of mixtures as compared to the pure fluids. The first is whether the pressure drop of mixtures is higher or lower than that of pure fluids. Torikoshi and Ebisu (1993) studied the pressure drop of R32/R134a (30/70 wt.%) and reported it follows an ideal pressure drop as predicted by the linear interpolation of the values of the pure fluids, as shown in Figure 8.4.1. This behavior is somewhat different from the boiling heat transfer behavior of mixtures as compared to that of pure fluids.

The second issue is whether the pressure drop of an alternative refrigerant mixture is higher or lower than that of the pure fluids it is supposed to replace. Wijaya and Spatz (1995) studied the pressure drop of R32/R125 (50/50 wt.%) during evaporation and found that R32/R125 results in approximately 24 to 38% lower pressure drop than that of R22.

8.5 CONDENSATION HEAT TRANSFER COEFFICIENTS

In this chapter, research on condensation heat transfer is summarized. The summary is limited to condensation in horizontal tubes since most of the condensers used for refrigeration and air-conditioning applications are oriented horizontally. Table 8.3 summarizes the latest research on the condensation heat transfer coefficients of

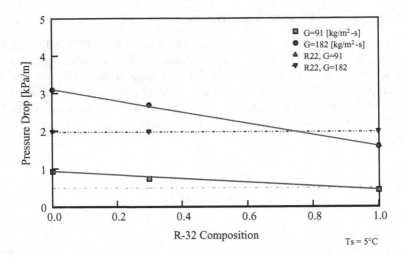

FIGURE 8.4.1 Pressure drop of R32/R134a during evaporation (Torikoshi and Ebisu, 1993).

TABLE 8.3
In-tube Condensation Research

Authors	Refrigerant	Mass Flux [kg/m²-s]	Tube/ID [mm]
Stoecker and Kornota (1985)	R12/R114	173	Plain/13
Tandon (1986)	R22/R12	170–530	Plain/10
Koyama et al. (1990)	R22/R114	133–358	Microfin/8.3
Conklin and Vineyard (1991)	R143a/R124	200–500	Finned/13.7, Fluted/16.8
Sami et al. (1992)	R22/R152a/R114, R22/R152a/R124	142–255 142–255	Fluted/21.2 Fluted/21.2
Sami et al. (1993)	R22/R114, R22/R152a	333–776	Microfin/8.3
Sami et al. (1995)	R23/R22/R152a		
Torikoshi and Ebisu (1993)	R32/R134a	100–400	Plain/8.7
Doerr et al. (1994)	R32/R125, R32/R134a, R32/R125/R134a, R125/R143a/R134a	125–375	Plain/OD 9.5
Ro et al. (1994)	R32/R134a	192	Plain/7.6
Dobson and Chato (1998)	R32/R125	25–800	Plain/3.14, 7.04
Wijaya and Spatz (1995)	R32/R125	160–561	Plain/7.8
Wang et al. (1996)	R407C	100–300	Plain/7.9
Eckels and Tesene (1999)	R410a, R407C	125–600	Plain/8.0 Microfin/8.9, 14.6, 7.3

Note: ID: Inside diameter, OD: Outside diameter.

Source: Wang and Chato, 1995.

mixtures based on Wang and Chato's (1995) summary and an additional survey of the latest literature. Wang and Chato pointed out that the main focus of the recent research on condensation heat transfer with mixtures is on:

- Thermal resistance of the vapor diffusion layer affecting the condensation.
- Influence of the flow direction of vapor during the condensation (horizontal tube or vertical tube).
- Turbulence in the vapor generated by fins.
- Enhancement of the condensation performance of mixtures.

During the mixture condensation process, the higher boiling component transfers into condensate at the vapor-liquid interface more so than the lower boiling component because in the condensate, the mass fraction of the high boiling point component is higher than in the vapor phase. More of the lower boiling point component diffuses into the vapor mixture away from the interface. This mass fraction gradient causes the deterioration of condensation heat transfer at the interface. Therefore, the greater volatility difference between the mixture components results in lowering the condensation heat transfer. This resistance to condensation heat transfer is similar to that of evaporation heat transfer. The resistance is seen as being caused by the originating phase.

Stoecker and Kornota (1985) investigated the condensation heat transfer coefficients of R12/R114. They divided the condenser into two parts, a high temperature condenser and a low temperature condenser. Their results can be summarized as follows:

- In the beginning of the high temperature condenser where the desuperheating occurs, no change of mixture mass fraction occurs and there is no effect on the condensing heat transfer coefficients.
- In the last part of the high temperature condenser where the phase changes, that is the midrange of the condenser, the condensing heat transfer coefficients are highest and have the most dominant changes. As the R114 mass fraction increases, the heat transfer coefficients decrease.
- Stocker and Kornota explained this heat transfer coefficient degradation by the slip between the liquid and vapor. No slip condition is assumed when modeling two-phase flow at low vapor velocities. However, when two-phase flow at high vapor velocities are considered, the vapor velocity at the liquid-vapor interface can exceed that of the liquid. This slip condition creates a local nonequilibrium state between the two phases with respect to temperature and mass fraction during the condensation of the mixtures.

Tandon et al. (1986) investigated the condensation heat transfer coefficients of R22/R12. They reported that the condensing heat transfer coefficients of mixtures are lower than those of R22 and higher than those of R12.

Koyama et al. (1990) investigated the condensing heat transfer coefficients of R22/R114 mixtures in a horizontal tube, then reported their results as follows:

- The local condensing heat transfer coefficients of the mixtures are smaller than those of pure refrigerants.
- The condensing heat transfer coefficients of a zeotropic mixture are enhanced in a microfin tube by about 30% as compared to a plain tube. This enhancement is only half as much compared to that of the pure refrigerant enhancement.
- The maximum degradation of the condensing heat transfer coefficient of R22/R114 is about 20% at 34 to 43 wt.% of R22 mass fraction.
- The circumferential wall temperature distribution is somewhat different from that of the pure refrigerants. For R22, the circumferential wall temperature variation is small at the entrance but gradually increases in the flow direction, indicating the flow pattern changes from the annular flow to semiannular flow. The circumferential wall temperature of R22 decreases from the right, top, bottom and left since the thin liquid film at the top moves to the right and the thick liquid film at the bottom moves to the left due to the internal groove. For R22/R114 mixtures, the circumferential wall temperature is uniform, which might be due to the opposite changes of liquid-vapor interfacial temperature distribution.

Conklin and Vineyard (1991) studied the condensing heat transfer coefficient of R143a/R124 (30/70, 51/49, 73/27 and 77/23 wt.%) in a finned tube and fluted tube. They reported that the condensing heat transfer coefficients of theses binary mixtures are higher in a finned tube than in a fluted tube. The condensing heat transfer coefficients of R143a/R124 (30/70 wt.%) are about 15% higher than that of R22. The pressure drop in the finned tube is only half of that in the fluted tube.

Sami et al. (1992) studied the condensing heat transfer coefficient of R22/R152a/R114 (53/19/28 wt.%) and R22/R152a/R124 (50/20/30 wt.%) in a fluted tube and reported the condensing heat transfer coefficient of these two ternary mixtures is lower than R22 and higher than R12. No significant difference was found between the condensing heat transfer coefficient of R22/R152a/R114 and R22/R152a/R124. By using the same facility, Sami and Schnotale (1993) studied the condensing heat transfer coefficient of R22/R114 and R22/R152a in a fluted tube. As the R114 mass fraction increases, the condensing heat transfer coefficient of R22/R114 increases almost linearly. However, the condensing heat transfer coefficient of R22/R152a decreases as the R152a mass fraction increases. They also found that the heat transfer coefficients of R22/R114 mixtures are lower than those of R22 but the heat transfer coefficients of R22/R152a mixtures are higher than those of R22.

Torikoshi and Ebisu (1993) studied the condensing heat transfer coefficient of R32/R134a (30/70 wt.%). They reported the degradation of the condensing heat transfer coefficient compared to the ideal condensing heat transfer coefficient as 36 and 29% for a mass flux of 91 kg/m^2-s and 182 kg/m^2-s, respectively, when the saturation temperature was 50°C.

TABLE 8.4
Ratio of In-tube Condensing Heat Transfer Coefficient to R22

Condensing Temperature	R32/R125 (60/40 wt.%)	R32/R134a (10/90 wt.%)	R32/R134a (25/75 wt.%)	R32/R125/R134a (30/10/60 wt.%)	R125/R143a/R134a (44/53/4 wt.%)
30°C	1.02	0.81	0.82	0.79	0.77
40°C	n/a	0.90	0.93	0.92	0.76

Mass flux: 200 kg/m²-s.

Source: Doerr et al., 1994.

Doerr et al. (1994) investigated the condensing heat transfer coefficients of R32/R125 (60/40 wt.%), R32/R134a (10/90 and 25/75 wt.%), R32/R125/R134a (30/10/60 wt.%) and R125/R143a/R134a (44/53/4 wt.%) in a horizontal tube and compared those with R22. As summarized in Table 8.4, only R32/R125 (60/40 wt.%) has a slightly higher heat transfer coefficient than R22. The other mixtures have a lower heat transfer coefficient than R22 by 4 to 7%.

Ro et al. (1994) investigated condensing heat transfer coefficient of R32/R134a (30/70 wt.%) and compared it with that of R22. Their results are:

- The heat transfer coefficient of R32/R134a (30/70 wt.%) is 10 to 20% lower than that of R22.
- The condensing heat transfer coefficient increases at lower vapor qualities as the vapor quality increases. However, the condensing heat transfer coefficient decreases at the higher vapor quality as the vapor quality increases. These characteristics are also represented by the existing correlations such as Chen (1966), Soliman et al. (1968) and Traviss (1973) that are based on the annular flow assumption. In the annular flow regime, the heat transfer coefficient decreases because the growing liquid layer around the tube increases thermal resistance.
- The condensing heat transfer coefficients of the mixture match better with these correlations at the higher vapor quality where the flow regime experiences annular flow. Most of the data fell in the annular flow regime. The mean deviation of theses correlations with the data is about 30%.

Wijaya and Spatz (1995) reported that the condensing heat transfer coefficient of R410A is slightly higher (about 3 to 6%) than that of R22 at low refrigerant quality since the liquid thermal conductivity of R410A is as much as 8% lower than that of R22 and its viscosity is 30 to 36% lower than that of R22.

Wang et al. (1996) compared the condensing heat transfer coefficient of R22 and R407C. They reported that the heat transfer coefficient of R407C is only 30 to 50% that of R22. The heat transfer coefficient of R407C is not very dependent upon the changes of condensing pressure though the heat transfer coefficient of R22 increases as the condensing pressure increases.

Eckels and Tesene (1999) compared the condensing heat transfer coefficients of R22, R134a, R410A and R407C in a smooth tube and microfin tubes. Their results are:

- The condensing heat transfer coefficients for the microfin tube are significantly higher than those for the smooth tube for all refrigerants tested.
- R134a has the highest heat transfer coefficients in both types of tubes while R407C has the lowest HTCs in both types of tubes.
- The heat transfer coefficient enhancement of the microfin tube over the smooth tube varies from 2.2 to 2.5 at the lowest mass flux and from 1.2 to 1.6 at the highest mass flux. However, only R407C has a lower enhancement at the lowest mass flux. This enhancement can be explained by the turbulence enhancement by the internal fins that delay the flow pattern transition from the annular flow to wavy flow.

8.6 CORRELATIONS FOR FLOW CONDENSATION HEAT TRANSFER

For modeling purposes, accurate heat transfer correlations for mixtures are needed to properly design heat exchangers. This section presents literature survey results from a screening of condensation heat transfer correlations.

In this section, four well-known condensing heat transfer correlations for pure fluids are introduced — Akers, Traviss, Shah and Dobson — and four recent correlations for mixtures — Tandon, Koyama, Sami and Sweeney — are explained. The applicability of these general correlations for mixtures must be proven.

8.6.1 AKERS CORRELATION

Akers et al. (1959) developed the following correlation (Equation 8.27) for the turbulent annual film flow that is only valid for pure fluids.

$$h = 0.0265\frac{k_l}{D}\text{Re}_{eq}^{0.8}\text{Pr}_l^{1/3}, \text{Re}_{eq} > 5\times10^4$$

$$h = 5.03\frac{k_l}{D}\text{Re}_{eq}^{0.8}\text{Pr}_l^{1/3}, \text{Re}_{eq} < 5\times10^4$$

(8.27)

where

Re_{eq} = Reynolds number at equivalent mass flux given by Equation 8.28

$$\text{Re}_{eq} = \text{Re}_{ls} + \text{Re}_{vs}\left(\frac{\mu_v}{\mu_l}\right)\left(\frac{\rho_l}{\rho_v}\right)^{0.5}$$

(8.28)

where
$$\mathrm{Re}_{ls} = GD\,(1-x)/\mu_l$$
$$\mathrm{Re}_{vs} = GDx/\mu_v$$

8.6.2 TRAVISS CORRELATION

Traviss et al. (1973) proposed the following semi-empirical correlation (Equation 8.29) based on their experimental data with R12 and R22 in a horizontal smooth tube.

$$h = \frac{k_l}{D}F(X_{tt})\frac{\mathrm{Re}_l^{0.9}\,\mathrm{Pr}_l}{F_2} \qquad (8.29)$$

where

$$F(X_{tt}) = 0.15\left|X_{tt}^{-1} + 2.85X_{tt}^{-0.476}\right|$$

$$X_{tt} = \left(\frac{\mu_l}{\mu_v}\right)^{0.1}\left(\frac{1-x}{x}\right)^{0.9}\left(\frac{\rho_v}{\rho_l}\right)^{0.5} \qquad (8.30)$$

$$F_2 = 0.707\,\mathrm{Pr}_l\,\mathrm{Re}_l^{0.5} \quad \text{for } \mathrm{Re}_l < 50$$

$$F_2 = 5\,\mathrm{Pr}_l + 5\ln\left[1 + \mathrm{Pr}_l(0.09636\,\mathrm{Re}_l^{0.585} - 1)\right] \quad \text{for } 50 < \mathrm{Re}_l < 1125 \qquad (8.31)$$

$$F_2 = 5\,\mathrm{Pr}_l + 5\ln(1 + 5\,\mathrm{Pr}_l) + 2.5\ln(0.00313\,\mathrm{Re}_l^{0.812}) \quad \text{for } \mathrm{Re}_l > 1125$$

8.6.3 SHAH CORRELATION

For predicting the heat transfer coefficients during film condensation inside pipes, Shah (1979) proposed an enhancement model that enhances the single-phase heat transfer coefficient of flowing liquid by a two-phase enhancement factor, as shown in Equation 8.32. This correlation was compared with many pure fluids such as water, R11, R12, R22, R113, methanol, ethanol, benzene, toluene and trichloroethylene. The mean deviation was 15.4%.

$$h = h_l\left[(1-x)^{0.8} + \frac{3.8x^{0.76}(1-x)^{0.04}}{\mathrm{Pr}^{0.38}}\right] \qquad (8.32)$$

where
h_l = single-phase liquid heat transfer coefficient by Dittus and Boelter equation

8.6.4 DOBSON CORRELATION

Dobson et al. (1994) developed their correlations for the annular- and wavy-flow regime (Equations 8.33 and 8.34, respectively) for R12 and R134a in a smooth tube that is of similar form as Dittus and Boelter's and Chato's correlations, respectively. They reported that these correlations match their data within a 7% deviation:

$$h = \frac{k_l}{D} 0.023 \, \mathrm{Re}_l^{0.8} \, \mathrm{Pr}_l^{0.4} \left(1 + 2.22 X_{tt}^{-0.89} \right) \text{ for annual-flow regime} \tag{8.33}$$

$$h = \frac{k_l}{D} \left[\frac{0.23 \, \mathrm{Re}_v^{0.12}}{1 + 1.11 X_{tt}^{0.58}} \left(\frac{Ga \, \mathrm{Pr}_l}{Ja_l} \right) + (1 - \theta_l / \pi) Nu_f \right] \text{ for wavy-flow regime} \tag{8.34}$$

where

$$Nu_f = 0.0195 \, \mathrm{Re}_l^{0.8} \, \mathrm{Pr}_l^{0.4} \, \phi_l(X_{tt})$$

where

$$\phi_l(X_{tt}) = \sqrt{1.376 + \frac{C_1}{X_{tt}^{c_2}}}$$

where

$$c_1 = 4.12 + 0.48 \mathrm{Fr}_l - 1.564 \mathrm{Fr}_l^2$$
$$c_2 = 4.12 + 0.48 \mathrm{Fr}_l - 1.564 \mathrm{Fr}_l^2 \text{ for } 0 < \mathrm{Fr}_l \le 0.7$$
$$c_1 = 7.242$$
$$c_2 = 1.655 \text{ for } \mathrm{Fr}_l > 0.7$$

where

Fr_l = liquid Froude number
θ_l = angle subtended from the top of tube to the liquid level

$$1 - \frac{\theta_l}{\pi} \cong \frac{\arccos(2\alpha - 1)}{\pi} \text{ by Jaster and Kosky (1976)}$$

8.6.5 TANDON CORRELATION

Tandon et al. (1986) proposed the following correlation (Equation 8.35) based on their experimental data with R22/R12 mixtures in a horizontal tube. The mean deviation was 15%:

$$h = 2.82 \frac{k_l}{D} Ph^{-0.365} \, \mathrm{Re}_v^{0.146} \, \mathrm{Pr}_l^{1/3} \tag{8.35}$$

where

$$Ph = \frac{h_{fg}}{c_{Pl}\left(T_{sat} - T_w\right)}$$

8.6.6 KOYAMA CORRELATION

Koyama et al. (1990) proposed the following correlation (Equation 8.36) based on their experimental data with R22/R114 mixtures in a horizontal microfin tube. Their correlation is a modified version of Fujii and Nagata's (1973).

$$h = 0.53\frac{k_l}{D}\left(\frac{D}{L}\right)^{0.4} Ph^{-0.6}\left(\frac{Re_l\, Pr_l}{R}\right)^{0.8} (1 - 0.73X + 0.37X^2 + 0.36X^3) \quad (8.36)$$

8.6.7 SAMI CORRELATION

Sami et al. (1992) developed their correlation (Equation 8.37) for mixtures of R22/R152a/R114 (53/19/28 wt.%) and R22/R152a/R124 (50/20/30 wt.%) in an enhanced tube that is of similar form as Koyama's correlation. They reported that this correlation matches with their data within a 10% deviation:

$$h = 6.4\frac{k_l}{D}\left(\frac{D}{L}\right)^{0.4} Ph^{-0.6}\left(\frac{Re_l\, Pr_l}{R}\right)^{0.6} \quad (8.37)$$

where

$$R = \left(\frac{\rho_l\mu_l}{\rho_v\mu_v}\right)^{0.5}$$

Later, Sami et al. (1995) reported that this correlation with a coefficient 0.212 instead of 6.4 matches with their experimental data for R23/R22/R134a (5/90/5 wt.%), R23/R22/R152a (5/90/5 wt.%), R23/R125/R152a (8/62/30 wt.%) and R23/R22/R134a (5/65/30 wt.%) within a 20% deviation.

8.6.8 SWEENEY CORRELATION

Sweeney (1996) and Sweeney and Chato (1996) correlated Kenney et al. (1994) data for R407C in a smooth tube using Dobson's correlations and suggested correlations, Equations 8.38 and 8.39:

$$h = 0.7\left(\frac{G}{300}\right)^{0.3} h_{Eq.\ (8.36)} \text{ for annual-flow regime} \quad (8.38)$$

$$h = \left(\frac{G}{300}\right)^{0.3} h_{Eq.\ (8.37)} \text{ for wavy-flow regime} \quad (8.39)$$

8.6.9 Comparison of Correlations

Judge (1996) compared five flow condensation correlations in his cycle simulation study to determine which one works best with the simulation to predict the overall heat transferred. The flow condensation heat transfer correlations examined in his study were: Tandon et al. (1986), Chen (1987), Traviss et al. (1973), Shah (1979) and Dobson et al. (1994). Dobson's correlation was developed for R32/R125 (50/50 wt.%) and Tandon's correlation was developed from R22/R12 data for several different mass fractions. The remaining correlations were developed for pure fluids.

The condensation results are presented in a similar format as the evaporation results and can be seen in Figure 8.6.1. Dobson's condensation correlation worked best with Judge's heat exchanger model; the maximum difference between the simulation and experimental results ranges from –3 to +7%. Typical condensation heat transfer coefficients vs. position in the heat exchanger are presented in Figures 8.6.2 and 8.6.3. The increased error in condensation relative to evaporation occurs for two reasons. One is that none of the condensation heat transfer correlations were developed to be generally applicable for any mixture. The other reason is that most of the correlations considered were developed for the annular flow regime.

The heat exchanger simulation was used to generate points on Baker's flow map (Baker, 1954). Figure 8.6.4 is a flow map of R22 in evaporation and condensation, respectively. For evaporation, a large fraction of the heat is absorbed in the annular regime. Furthermore, the other region representing wavy flow is somewhat similar to annular flow in that the entire circumference of the tube can be covered with liquid. Hence, annular flow correlations are often used to approximate wavy flow. In any event, Jung's flow boiling correlation is not restricted to the annular flow regime. At low mass fluxes, Jung's correlation reduces to pool boiling, an attractive aspect since the pool boiling represents the lower limit for all other boiling regimes. For condensation, the flow map shows that a significant portion of the heat is rejected in the slug and plug flow regimes. This is unfortunate since there exists very little heat transfer data in these regimes. However, Dobson's condensation correlation for the annular flow regime reduces to a single-phase flow correlation as the quality is reduced. Hence, Dobson's correlation is accurate in the limits of single-phase flow and annular flow and it is used to approximate the heat transfer coefficient for the flow regimes between them. Furthermore, the sample size is not sufficiently large to state that the heat exchanger model will predict the capacity for any heat exchanger configuration within the errors reported here.

8.7 PRESSURE DROP DURING CONDENSATION

In this section, recent experimental research and correlations for horizontal condensing flow pressure drop of mixtures are summarized. As mentioned in Section 8.4, the pressure drop in a pipe consists of three major contributions: frictional pressure drop (ΔP_f), accelerational pressure drop (ΔP_a) and gravitational pressure drop (ΔP_g). For the horizontal tube, the gravitational pressure drop is eliminated ($\Delta P_g = 0$).

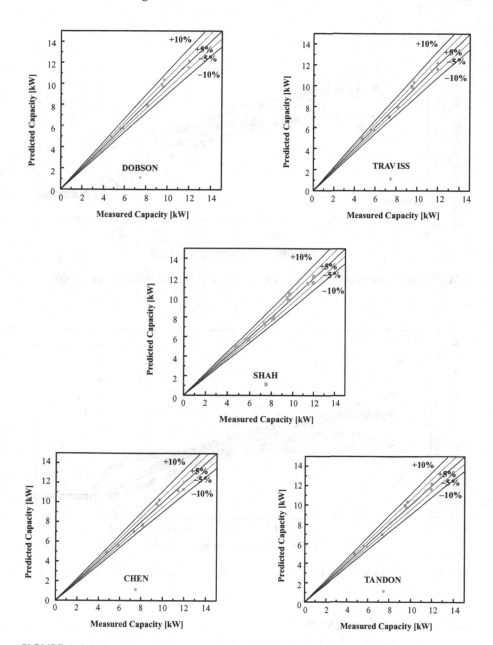

FIGURE 8.6.1 Condensation heat transfer correlation results (Judge, 1996).

Moreover, the accelerational pressure drop is generally much smaller than the frictional pressure drop; therefore, most investigators only considered the frictional pressure drop. Two major parameters of the frictional pressure drop are the liquid viscosity and mass flux.

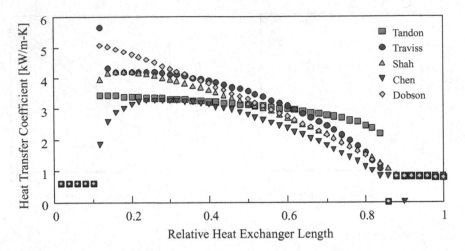

FIGURE 8.6.2 Condensation heat transfer coefficient vs. length (R22) (Judge, 1996).

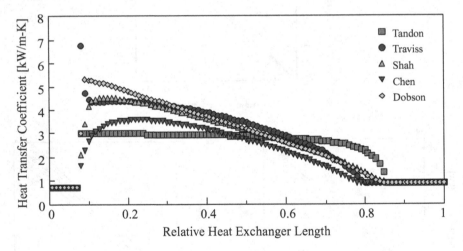

FIGURE 8.6.3 Condensation heat transfer coefficient vs. length (R407C) (Judge, 1996).

8.7.1 KOYAMA CORRELATION

Koyama et al. (1990) suggested the following correlation based on their experiment with R22/R114 mixtures in a horizontal microfin tube. They reported their results for frictional pressure drop are well-correlated by the Lockhart and Martinelli parameters within a 30% deviation, as shown in Equations 8.40 through 8.44:

$$\Delta P = \Delta P_f + \Delta P_m \tag{8.40}$$

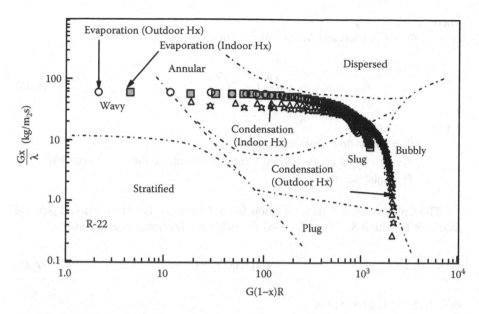

FIGURE 8.6.4 Baker's flow map: Conditions for evaporation and condensation processes are superimposed (Judge, 1996).

where

ΔP = total two-phase pressure drop
ΔP_f = frictional pressure drop
ΔP_m = pressure drop due to the momentum change

$$\Delta P_m = \frac{G^2}{\rho_l}\left(\left[\frac{x^2}{\alpha}\left(\frac{\rho_l}{\rho_v}\right)+\frac{(1-x)^2}{(1-\alpha)}\right]_2 - \left[\frac{x^2}{\alpha}\left(\frac{\rho_l}{\rho_v}\right)+\frac{(1-x)^2}{(1-\alpha)}\right]_1\right) \quad (8.41)$$

where

G = mass flux
ρ_l = liquid density
ρ_v = vapor density
x = vapor quality
α = void friction

$$\alpha = \left\{1+\left(\frac{\rho_v}{\rho_l}\right)\left(\frac{1-x}{x}\right)\left(0.4+0.6\left[\frac{x\left(\rho_l/\rho_v\right)+0.4(1-x)}{x+0.4(1-x)}\right]^{0.5}\right)\right\}^{-1} \quad (8.42)$$

$$\Delta P_f/\Delta z = \left(\Delta P/\Delta z\right)_v \Phi_v^2 \quad (8.43)$$

where

Φ_v = Lockhart and Martinelli parameter

$$\left(\frac{\Delta P}{\Delta z}\right)_v = \frac{2f_v G^2 x^2}{D\rho_v} \tag{8.44}$$

where

ΔP_v = vapor flow pressure drop
f_v = frictional coefficient considering two-phase flow as vapor flow
D = tube diameter

The Colburn friction factor relation for the turbulent flow in a smooth pipe (as shown in Equation 8.45) can be used to obtain the frictional coefficient (f_v),

$$f_v = 0.046\,\text{Re}^{-0.2} \tag{8.45}$$

8.7.2 SAMI CORRELATION

Sami et al. (1992) suggested using the Lockhart and Martinelli parameters to calculate the frictional pressure gradient based on their measurement of R22/R152a/R114 and R22/R152a/R124 in an enhanced tube. The suggested correlation is shown in Equations 8.40 and 8.41. They correlated the friction factor using the form of Blasius as shown in Equation 8.46:

$$f = 0.6\,\text{Re}^{-0.21} \tag{8.46}$$

The Sami correlation matches their experimental data within 30%. Later, Sami et al. (1995) examined their earlier correlation against their data for other mixtures of R23/R22/R152a (5/90/5 and 5/65/30 wt.%), R23/R22/R290 (5/90/5 wt.%), R23/R22/R134a (5/65/30 wt.%) and R23/R125/R152a (8/62/30 wt.%) and reported that the mean deviation is 15% for most of the data.

8.7.3 OTHER STUDIES

There are two interesting issues regarding pressure drop of mixtures as compared to the pure fluids as mentioned in Section 8.4.6. The first is whether the pressure drop of mixtures is higher or lower than that of pure fluid constituents. Torikoshi and Ebisu (1993) studied the pressure drop of R32/R134a (30/70 wt.%) and reported it follows an ideal pressure drop as predicted by the linear interpolation of the values of the pure fluids. However, the pressure drop of R32/R134a (30/70 wt.%) is about 20% larger than that of R22. This behavior is somewhat different from the condensing heat transfer behavior of mixtures as compared to that of pure fluids.

The second issue is whether the pressure drop of an alternative refrigerant mixture is higher or lower than those of the pure fluids it is intended to replace. Wijaya and Spatz (1995) studied the pressure drop of R32/R125 (50/50 wt.%) during condensation and found that R32/R125 results in an approximately 25 to 45% lower pressure drop than that of R22.

REFERENCES

Akers, W.W., H.A. Deans and O.K. Crosser, 1959, "Condensing Heat Transfer within Horizontal Tubes," Chemical Engineering Progress Symposium series, Vol. 55, No. 29, pp. 171–176.

Baker, O., 1954, "Design of Pipe Lines for Simultaneous Flow of Oil and Gas," *Journal of Oil and Gas*, July, No. 26.

Bennett, D.L. and J.C. Chen, 1980, "Forced Convective Boiling in Vertical Tubes for Saturated Pure Components and Binary Mixtures," *Journal of AIChE*, Vol. 26, pp. 454–461.

Bivens, D.B. and A. Yokozeki, 1994, Proceedings of 1994 International Refrigeration Conference at Purdue, West Lafayette, IN.

Bivens, D.B. and A. Yokozeki, 1994, *A Heat Transfer of Zeotropic Refrigerant Mixtures Heat Pumps for Energy Efficiency and Environmental Progress,* Elsevier, Amsterdam.

Buschmeier, M., P. Sokol, A.D. Pinto, and D. Gorenflo, 1994, "Pool Boiling Heat Transfer of Propane/n-Butane Mixtures at a Single Tube with Superimposed Convective Flow of Bubbles or Liquid," Proceedings of 10th International Heat Transfer Conerence, Vol. 5, pp. 69 74.

Chen, J.C., 1966, "A Correlation for Boiling Heat Transfer to Saturated Fluids in Vertical Flow," *Ind. Eng. Chem. Proc. Design Dev.*, Vol. 5, No. 3, pp. 322–339.

Chen, S.L., F.M. Gerner, and C.L. Tein, 1987, "General Film Condensation Correlations," *Experimental Heat Transfer,* Vol. 1, pp. 93–107.

Chisholm, D., 1967, "A Theoretical Basis for the Lockhart-Martinelli Correlation for Two-Phase Flow," *International Journal of Heat and Mass Transfer*, Vol. 10, pp. 1767–1778.

Chisholm, D., 1973, "Pressure Gradients Due to Friction During the Flow of Evaporating Two-phase Mixtures in Smooth Tubes and Channels," *International Journal of Heat and Mass Transfer*, Vol. 16, pp. 347–358.

Collier, J.G. and J.R. Thome, 1994, *Convective Boiling and Condensation*, Clarendon Press Oxford, England, pp. 535–584.

Conklin, J.C. and E.A. Vineyard, 1990, "Tube-side Evaporation of Non-Azeotropic Refrigerant Mixtures from Two Enhanced Surfaces," ASME Winter Meeting.

Conklin, J.C. and E.A. Vineyard, 1991, "Tube-side Condensation of Non-Azeotropic Refrigerant Mixtures for Two Enhanced Surfaces," ASME/JSME Thermal Engineering Proceedings, Vol. 3, pp. 251–256.

Cooper, M.G., 1984, "Saturation Nucleate Pool Boiling, A Simple Correlation," 1st U.K. National Conference on Heat Transfer, Vol. 2, pp. 785–793.

Darabi, J., M. Salehi, M.H. Saeedi and M.M. Ohadi, 1995, "Review of Available Correlations for Prediction of Flow Boiling Heat Transfer in Smooth and Augmented Tubes," *ASHRAE Transactions*, Vol. 101, Pt. 1, pp. 965–975.

Dittus, F.W. and L.M.K. Boelter, 1930, *University of California Publications on Engineering*, Vol. 2, p. 443.

Dobson, M.K., J.C. Chato, D.K. Hinde and S.P. Wang, 1994, "Experimental Evaluation of Internal Condensation of Refrigerants R12 and R134A," *ASHRAE Transactions*, Vol. 100, Pt. 1, pp. 744–754.

Dobson, M.K. and J.C. Chato, 1998, "Condensation in Smooth Horizontal Tubes," *Journal of Heat Transfer*, Vol. 120, pp. 193–213.

Doerr, T. and M.B. Pate, 1993, "In-Tube Evaporation and Condensation Alternatives for R22 and R502," ASHRAE/NIST Refrigerants Conference, R22/R502 Alternative Session II, Gaithersburg, MD.

Doerr, T.M., S.J. Eckels and M.B. Pate, 1994, "In-Tube Condensation Heat Transfer of Refrigerants Mixtures," *ASHRAE Transactions*, Vol. 100, Pt. 2, pp. 547–557.

Eckels, S.J. and M.B. Pate, 1991, "In-tube Evaporation and Condensation of a Ternary Refrigerant Blend and HFC-134a," ERI Project 3315, Iowa State University.

Eckels, S.J. and B.A. Tesene, 1999, "A Comparison of R22, R134A, R410A, and R407C Condensation Performance in Smooth and Enhanced Tubes: Part I, Heat Transfer," *ASHRAE Transactions*, Vol. 105, Pt. 2.

Foster, H.K. and N. Zuber, 1955, *Journal of AIChE*, Vol. 1, pp. 531–535.

Fujii, T., T. Nagata, 1973, "Condensation Vapor in a Horizontal Tube," Report of Research Institute of Industrial Science, Kyushu University, Vol. 52, pp. 35–50.

Gorenflo, D., P. Kaupman, R. Koster, and M. Buschmeier, 1995, "Pool Boiling Heat Transfer of Propane/i-Butane/n-Butane Mixtures," Proceedings of the 19th International Congress of Refrigeration, IVa, pp. 238–245.

Goto, M., N. Inoue, and K. Koyama, 1995, "Evaporation Heat Transfer of HCFC22 and Its Alternative Refrigerants inside an Internally Grooved Horizontal Rube," Proceedings 195th International Congress of Refrigeration, IVa, pp. 246–253.

Gungor, K.E. and R.H.S. Winterton, 1986, *International Journal of Heat and Mass Transfer*, Vol. 29, No. 3, pp. 351–358.

Hihara, E., K. Tanida, and T. Saito, "1989, Forced Convective Boiling Experiments of Binary Mixtures," *JSME International Journal,* Series II, Vol. 32, No. 1, pp. 98–106.

Jaster, H. and P.G. Kosky, 1976, "Condensation in a Mixed Flow Regime," *International Journal of Heat and Mass Transfer*, Vol. 19, pp. 95–99.

Judge, J., 1996, "A Transient and Steady State Study of Pure and Mixture Refrigerants in a Residential Heat Pump," Ph.D. Dissertation, University of Maryland, College Park, MD.

Judge, J. and R. Radermacher, 1997, "A Heat Exchanger Model for Mixtures and Pure Refrigerant Cycle Simulation," *International Journal of Refrigeration*, Vol. 20, No. 4, pp. 244–255.

Jung, D.S. and R. Radermacher, 1989, Prediction of Pressure Drop during Horizontal Annular Flow Boiling of Pure and Mixed Refrigerants, *International Journal Heat Mass Transfer,* Vol. 32, pp. 2435–2446.

Jung, D.S., M. McLinden, R. Radermacher and D.A. Didion, 1989a, "A Study of Flow Boiling Heat Transfer with Refrigerant Mixtures," *International Journal of Heat Transfer*, Vol. 32, No. 9, pp. 1751–1764.

Jung, D.S., M. McLinden, R. Radermacher and D.A. Didion, 1989b, "Horizontal Flow Boiling Heat Transfer Experiments with a Mixture of R22/R114," *International Journal of Heat Transfer*, Vol. 32, No. 1, pp. 131–145.

Kattan, N., J.R. Thome, and D. Favrat, 1994, "In Tube Flow Boiling of R-502 and Two Near-Azeotropic Alternatives," Proceedings 10th International Heat Transfer Conference, Brighton, Vol. 7, pp. 455–460.

Kenney, P.J., R.L. Shinon, T.C. Villanueva, N.L. Rhines, K.A. Sweeney, D.G. Allen, and T.T. Hershberger, 1994, "Heat Transfer Flow Regimes of Refrigerants in a Horizontal Tube Evaporator, ACRC Technical Report 55, Univ. of Illinois.

Koyama, S., A. Miyara, H. Takamatsu and T. Fujii, 1990, "Condensation Heat Transfer of Binary Refrigerant Mixtures of R22 and R114 Inside a Horizontal Tube with Internal Spiral Grooves," *International Journal of Refrigeration*, Vol. 13, pp. 256–263.

Kutateladze, S.S., 1961, "Boiling Heat Transfer," *International Journal of Heat and Mass Transfer*, Vol. 4, pp. 31–45.

Liu, Z. and R.H.S. Winterton, 1991, "A General Correlation for Saturated and Subcooled Flow Boiling in Tubes and Annuli Based on a Nucleate Pool Boiling," *International Journal of Heat and Mass Transfer*, Vol. 34, pp. 2759–2765.

Martinelli, R.C. and Nelson, D.B., 1948, "Prediction of Pressure Drop During Forced Circulation Boiling of Water," *Transactions of ASME*, Vol. 70, pp. 695–702.

Melin, P. and L. Vamling, 1995, "Flow Boiling Heat Transfer and Pressure Drop for Zeotropic Mixtures in a Horizontal Tube, Two-Phase Flow Modeling and Experimentation," Vol. 1, pp. 609–616.

Mulroy, W., M. Kaulfeld, M. McLinden, and D. Didion, "An Evaluation of Two Refrigerant Mixtures in a Breadboard Air Conditioner Status of CRCs — Refrigerant Systems and Refrigerant Properties," IIR, Paris, pp. 27–34.

Murata, K. and K. Hashizume, 1990, "An Investigation on Forced Convection Boiling of Nonazeotropic Refrigerant Mixtures, *Heat Transfer-Japanese Research*, Vol. 19, No. 2, pp. 95–109.

Murata, K. and K. Hashizume, 1993, "Forced Convective Goiling of Nonazeotropic Refrigerant Mixtures Inside Tubes," *Journal Heat Transfer*, Vol. 115, pp. 680–689.

Niederkruger, M., D. Steiner, and E.U. Sehliinder, 1992, "Horizontal Flow Boiling Experiments of Saturated Pure Components and Mixtures of R846-R12 at High Pressures," *International Journal Refrigeration*, Vol. 15, No. 1, pp. 48–58.

Ohata, H. and Y. Fujta, 1994, "Nucleate Pool Boiling of Binary Mixtures," Proceedings of the 10th International Heat Transfer Conference, Vol. 5, pp. 129–134.

Pierre, B., 1956, "Coefficient of Heat Transfer for Boiling Freon-12," *Horizontal Tubes, Heating and Air Treatment Engineer*, Vol. 19, pp. 302–310.

Radermacher, R., H. Ross, and D. Didion, 1983, "Experimental Determination of Forced Convective Evaporative Heat Transfer Coefficients for Non-azeotropic Refrigerant Mixtures," ASME Paper, 83-WA/HT54.

Ro, S.T., H.S. Jun, Y.S. Chang and J.Y. Shin, 1994, "Condensation Heat Transfer of a Heat Pump System using the Refrigerant Mixtures R32/R134a," *ASHRAE Transactions*, Vol. 100, Pt. 2, pp. 715–734.

Rohlin, P., 1994, "Heat Transfer Coefficients in Horizontal Flow Boiling of Some Pure Refrigerants and Their Zeotropic Mixtures: Experimental and Theoretical Results," Proceedings of IIR Conference, CFCs, The Day After, pp. 583–590.

Ross, H., R. Radermacher, M. di Marzo, and D. Didion, 1987, "Horizontal Flow Boiling of Pure and Mixed Refrigerants," *Int. J. Heat Mass Transfer*, Vol. 30, pp. 979–992.

Sami, S.M. and T.N. Duong, 1991, "Experimental Study of the Heat Transfer Characteristics of Refrigerant Mixture R-22/R-114 in the Annulus of Enhanced Surface Tubing," *International Communication Heat Mass Transfer*, Vol. 18, pp. 547–558.

Sami, S.M. and T.N. Duong, 1991, "Flow Boiling Characteristics of Refrigerant Mixture R-22/R-114 in the Annulus of Enhanced Surface Tubing Heal Recovery Systems & CHP," Vol. 12, No. 1, pp. 39–48.

Sami, S.M. and J. Scbnotale, 1992, "Comparative Study of Two Phase Flow Boiling of Refrigerant Mixtures and Pure Refrigerants Inside Enhanced Surface Tubing," *International Communication Heat Mass Transfer*, Vol. 19, pp. 137–148.

Sami, S.M., J. Schnotale and J.G. Smale, 1992, "Prediction of the Heat Transfer Character-istics of R22/R152a/R114 and R22/R152a/R124," *ASHRAE Transactions*, Vol. 98, Pt. 2, pp. 51–58.

Sami, S.M. and J. Schnotale, 1993, "Prediction of Forced Convective Condensation of Non-Azeotropic Refrigerant Mixtures Inside Enhanced Surface Tubing," *Applied Scientific Research*, Vol. 50, pp. 149–168.

Sami, S.M., P.J. Tulej and B. Song, 1995, "Study of Heat and Mass Transfer Characteristics of Ternary Nonazeotropic Refrigerant Mixtures Inside Air/Refrigerant-Enhanced Sur-face Tubing," *ASHRAE Transactions*, Vol. 101, Pt. 1, CH-95-23-4.

Schmidt, J.G. 1995, Blasenbildung beim siedon von stoffgemischen, Ph.D. Thesis, University of Karlsruhe.

Schrock, V.E. and L.M. Grossman, 1959, "Forced Convective Boiling Studies," University of California, Institute of Engineering Research, Report No. 73308-UCX-2182, Berke-ley, California.

Shah, M.M., 1976, "A New Correlation for Heat Transfer During Boiling Flow through Pipes," *ASHRAE Transactions*, Vol. 82, No. 2, pp. 185–196.

Shah, M.M., 1979, "A General Correlation for Heat Transfer During Film Condensation Inside Pipes," *Journal of Heat and Mass Transfer*, Vol. 22, pp. 547–556.

Shah, M.M., 1982, "Chart Correlation for Saturated Boiling Heat Transfer: Equations and Further Studies," *ASHRAE Transactions*, Vol. 88, Pt. 1, pp. 185–196.

Shin, J.Y., M.S. Kim and S.T. Ro, 1996, "Correlation of Evaporative Heat Transfer Coefficients for Refrigerant Mixtures," Proceedings of the 1996 International Refrigeration Con-ference at Purdue, West Lafayette, IN, pp. 151–156.

Shin, J.Y., M.S. Kim and S.T. Ro, 1997, *International Journal of Refrigeration*, Vol. 20, No. 4, pp. 267–275.

Singal, L.C., C.P. Sharma and H.K. Varma, 1983a, "Pressure Drop During Forced Convection Boiling of Binary Refrigerant Mixtures," *International Journal of Multiphase Flow*, Vol. 9, No. 3, pp. 309–323.

Singal, L.C., C.P. Sharma, H.K. Varma, 1983b, "Experimental Heat ransfer Coefficient for Binary Refrigerant Mixtures of R13 and R12," *ASHRAE Transactions*, pp. 165–188.

Soliman, H.M., J.R. Schuster, and P.J. Berenson, 1968, "A General Heat Transfer Correlation for Annular Flow Condensation," *ASME Journal Heat Transfer*, Vol. 90, pp. 267–276.

Souza, A.L. and M.M. Pimenta, 1995, "Prediction of Pressure Drop During Horizontal Two-phase Flow of Pure and Mixed Refrigerants," ASME Conference on Cavitation and Multiphase Flow, FED-Vol. 210, pp. 161–171.

Stephan, K. and P. Preusser, 1979, "Heat Transfer and Critical Heat Flux in Pool Boiling of Binary and Ternary Mixtures," *German Chemical Engineering*, Vol. 2, pp. 161–169.

Stephen, K. and M. Abdelsalam, 1981, "Heat Transfer Correlations for Natural Convection Boiling," *International Journal Heat Mass Transfer*, Vol. 23, pp. 73–87.

Sternling, C.V. and L.J. Tichacek, 1961, "Heat Transfer Coefficients for Boiling Mixtures," *Chemical Engineering Science*, Vol. 16, pp. 297–337.

Stoecker, W.F. and E. Kornota, 1985, "Condensing Coefficients When Using Refrigerant Mixtures," *ASHRAE Transactions*, Vol. 91, Pt. 2B, pp. 1351–1367.

Sundaresan, S.G., M.B. Pate, T.M. Doerr and D.T. Ray, 1996, Proceedings of 1996 Interna-tional Refrigeration Conference at Purdue, West Lafayette, IN, pp. 187–192.

Sweeny, K.A., 1996, "The Heat Transfer and Pressure Drop Behavior of a Zeotropic Refrig-erant Mixture in a Microfin Tube," M.S. Thesis, University of Illinois at Urbana-Champaign.

Sweeny, K.A. and J.C. Chato, 1996, "The Heat Transfer and Pressure Drop Behavior of a Zeotropic Refrigerant Mixture in a Microfin Tube," ACRC Technical Report 95, University of Illinois at Urbana-Champaign.

Tandon, T.N., H.K. Varma and C.P. Gupta, 1986, "Generalized Correlation for Condensation of Binary Mixtures Inside a Horizontal Tube," *International Journal of Refrigeration*, Vol. 9, pp. 134–136.

Thome, J.R., 1982, "Latent and Sensible Heat Transfer Rates in the Boiling of Binary Mixture," *Journal of Heat Transfer*, Vol. 104, pp. 474–478.

Thome, J.R., 1983, "Prediction of Binary Mixture Boiling Heat Transfer Coefficients Using Only Equilibrium Data," *International Journal of Heat and Mass Transfer*, Vol. 26, No. 7, pp. 965–974.

Thome, J.R., 1996, "Boiling of New Refrigerants: A State-of-Art Review," *International Journal of Refrigeration*, Vol. 19, No. 7, pp. 435–457.

Torikoshi, K and T. Ebisu, 1993, "Heat Transfer and Pressure Drop Characteristics of R-134a, R-32, and a Mixture of R-32/R134a Inside a Horizontal Tube," *ASHRAE Transactions*, Vol. 99, Pt. 2, pp. 90–96.

Torikoshi, K. and T. Ebisu, 1993b, "In-Tube Heat Transfer and Characteristics of Azeotropic Mixtures of HFC-32 and HFC-125," ASHRAE/NIST Refrigerants Conference, R22/R502 Alternatives Session II, Gaithersburg, MD.

Traviss, D.P., W.M. Rohsenow and A.B. Baron, 1973, "Forced-Convection Condensation Inside Tubes: A Heat Transfer Equation for Condenser Design," *ASHRAE Transactions*, Vol. 79, Pt. 1, pp. 157–165.

Trewin, R.r., M.K. Jensen, and A.E. Bergles, 1994, "Pool Boiling from Enhanced Surfaces in Pure and Binary Mixtures of R113 and R11," Proceedings of 10th International Heat Transfer Conference, Vol. 5, pp. 165–170.

Van Wijk, W.R., A.S. Vos and S.J.D. Van Stralen, 1956, "Heat Transfer to Boiling Binary Liquid Mixtures," *Chemical Engineering Science*, Vol. 5, pp. 68–80.

Wallis, G.B., 1969, *One-Dimensional Two-Phase Flow*, McGraw-Hill, New York, NY, pp. 29–58.

Wang, C.C., C.S. Kuo, Y.J. Change and D.C. Lu, 1996, "Two-Phase Flow Heat Transfer and Friction Characteristrics of R22 and R407C," *ASHRAE Transactions*, Vol. 102, Pt. 1, pp. 830–838.

Wang, S.P. and J.C. Chato, 1995a, "Review of Recent Research on Heat Transfer with Mixtures-Part 1: Condensation," *ASHRAE Transactions*, Vol. 101, Pt. 1, pp. 1376–1386.

Wang, S.P. and J.C. Chato, 1995b, "Review of Recent Research on Heat Transfer with Mixtures-Part 2: Boiling and Evaporation," *ASHRAE Transactions*, Vol. 101, Pt. 1, pp. 1387–1401.

Wattelet, J.P., J.C. Chato, A.L. Souza, and B.R. Christofferson, 1993, Initial Evaporative Comparison of R-22 with Alternative Refrigerant R134a and R-32/R-125, Air Conditioning and Refrigeration Center Report TR-39, University of Illinois.

Uchida, M., M. Itoh, N. Shikazono and M. Kudoh, 1996, "Experimental Study on the Heat Transfer Performance of a Zeotropic Refrigerant Mixture in Horizontal Tubes," Proceedings of the 1996 International Refrigeration Conference at Purdue, West Lafayette, IN, pp. 133–138.

Wattelet, J.P., J.C. Chato, B.R. Christoffersen, J.A. Gaibel, M. Ponchner, P.J. Kenny, R.L. Shimon, T.C. Villanueva, N.L. Rhines, K.A. Sweeny, D.G. Allen and T.T. Hershberger, 1994a, "Heat Transfer Flow Regimes of Refrigerants in a Horizontal-Tube Evaporator," ACRC Report TR-55.

Wattelet, J.P., J.C. Chato, A.L. Souza and B.R. Christoffersen, 1994b, "Evaporative Characteristics of R12, R134a, and a Mixture at Low Mass Fluxes," *ASHRAE Transactions*,
Vol. 100, Pt. 1, pp. 603–615.

Webb, R.L. and N.S. Gupte, 1992, "A Critical Review of Correlations for Convective Vaporization in Tubes and Tube Banks," *Heat Transfer Engineering*, Vol. 13, No. 3, pp.
58–81.

Whalley, P.B., 1994, *Boiling, Condensation, and Gas-liquid Flow*, Clarendon Press Oxford,
England, pp. 167–173.

Wijaya, H. and M.W. Spatz, 1995, "Two-Phase Flow Heat Transfer and Pressure Drop
Characteristics," *ASHRAE Transactions*, Vol. 101, Pt. 1, pp. 1020–1030.

9 Operational Issues

This chapter discusses operational and handling issues that are unique to refrigerant mixtures. It is based on laboratory experience with refrigerant mixtures and guidelines by the U.S. Environmental Protection Agency (EPA) and the Air-Conditioning and Refrigeration Institute (ARI) for the handling of refrigerants.

9.1 REGULATION

9.1.1 CLEAN AIR ACT AMENDMENTS OF 1990

In 1990, the Clean Air Act was amended to control the production and use of refrigerants in the United States. The goal of these amendments was to reduce the total use and emissions of chlorofluorocarbons (CFCs) and hydrochlorofluorocarbons (HCFCs) and to maximize recycling. Two pertinent sections to the Clean Air Act, Sections 608 and 609, restrict the emissions of class I (CFCs) and II (HCFCs) substances from service, repair and disposal operations on stationary equipment and mobile applications, respectively. It is unlawful in the United States for any person to knowingly vent, release or dispose of any class I and II substances used as a refrigerant in appliances and industrial process refrigeration while maintaining, servicing, repairing or disposing, effective July 1, 1992. The EPA has the authority to write necessary regulations to implement the Clean Air Act.

9.1.2 EPA REGULATIONS

In 1993, the EPA set its regulations to enforce the Clean Air Act, which covers the following five sections:

- Require technicians servicing and disposing of air-conditioning and refrigeration equipment to observe service practices that reduce refrigerants emissions.
- Establish air-conditioning and refrigeration technician certification programs and restriction of the sale of refrigerants to certified technicians.
- Establish equipment and reclaimer certification programs.
- Require repair of substantial leaks.
- Implement safe disposal procedures.

Violations of the Clean Air Act are punishable by fines up to $25,000 per violation per day and the possible loss of technician certification. However, the EPA allows some minimal emissions in the course of handling of refrigerants and operation of equipment in the following cases:

- *De minimis* release if they occur, though all requirements for the proper handling of refrigerants of allowed and approved recovery and recycling machines are used.
- Refrigerant release from the equipment while in the normal operation, such as purging and leaks.
- Nitrogen mixed with a small mass fraction of R22 for leak testing.
- Refrigerant release from hose purging or connections while charging or servicing.

9.2 REFRIGERANT RECOVERY

First, it is necessary to define three terms for the handling of refrigerants that have already been used. These are well defined by an ARI (1994) guideline:

1. *Recovery*: the transferring of refrigerants from a system or container to a recovery container.
2. *Recycling*: the cleaning of refrigerants by oil separation, noncondensable removal and reduction of moisture, acidity and particulate matter for reuse.
3. *Reclaiming*: a complicated distillation process that returns recovered refrigerants to the purity of factory-new refrigerants.

Figure 9.2.1 shows a typical refrigerant recovery machine. Refrigerant recovery needs to be done very carefully to minimize refrigerant emissions, as can be seen in the following procedures:

- Prepare a recovery container that is approved by the Department of Transportation (DOT) for recovery purposes. Figure 9.2.2 shows a typical 9 kg recovery container.
- Evacuate the recovery container using a vacuum pump.
- Decide on a recovery method, either vapor recovery or liquid recovery, according to feasibility. Vapor recovery is slow (0.5 to 0.9 kg/min) but can remove most of the refrigerant from a system. While the liquid recovery is fast, it cannot remove the refrigerant vapor remaining in the system. For high pressure systems, vapor recovery is recommended.
- Before connecting any hoses to the system, make sure the system is off-duty throughout the recovery process. If any (solenoid) valves are used in the system, open the valves to enhance the recovery if possible.
- Connect the system service port, recovery container and recovery machine with refrigerant hoses.
- Open valves between the system, recovery machine and recovery container. Start the recovery machine.
- When the recovery container is filled to 80% of maximum (usually determined by weight), it should be disconnected and a new container should be used.

FIGURE 9.2.1 Refrigerant recovery machine. (Photo courtesy of INFICON.)

FIGURE 9.2.2 Recovery container. (Photo courtesy of Worthington Cylinder Corporation.)

9.3 CYCLE FLUSHING

When the system is retrofitted with another type of refrigerant and oil, the system should be flushed. It is necessary to check the types of refrigerants and oils used before and after the retrofit procedure. CFCs and HCFCs have mostly used mineral oils; however, hydrofluorocarbons (HFCs) as alternatives for CFCs and HCFCs require POE (Polyolester) oils. Many different types of POEs are available but some compressor and system manufactures recommend only specific oils based upon their own compressor life testing results. Therefore, special care must be taken to choose appropriate oils for the refrigerant used for each specific system. It is important for handling POEs to minimize the contact time with air since POEs are highly hydroscopic and therefore any opportunity to accummulate moisture must be avoided. As an example, a cycle flushing procedure, when the system originally charged with R22 is retrofitted with R407C, is explained as follows:

- Check refrigerant and oil compatibility: R22 and mineral oils are compatible, as are R407C and POE oils. However, R407C is not compatible with mineral oils. Therefore, a flushing of the system is needed.
- Old refrigerant removal: Recover old refrigerant charged in the system using a recovery machine.
- Initial oil drain from the compressor: Drain all old oil from the compressor through the drain port of the compressor, if any is available. If no oil drain port is available, separate the compressor from the system and drain the oil from the compressor through the ports.
- Initial oil drain from the heat exchangers: Drain as much of the oil from the cycle by blowing off with compressed nitrogen gas through the service ports.
- Reassemble the compressor if it was disassembled.
- Evacuate the system using a vacuum pump.
- Fill the system with new lubricant (POE) to the minimum quantity required and charge the system with R407C (use recovered R407C if available).
- Run the system for approximately 10 minutes to help any residual old refrigerant gas and oil to mix with new refrigerant and oil.
- Repeat above procedures except initial oil removal from the heat exchangers until the refractive index difference between the oil sample and new oil is within 1%.

9.4 REFRIGERANT MIXING

This section describes a generalized procedure for blending refrigerants. As an example, the blending of R32 and R125 to make a binary mixture R410A is explained. It should be noted that special care must be taken when blending flammable refrigerants such as R32, R152a and hydrocarbons.

- Prepare two containers charged with pure components, R32 and R125, and one empty container for the mixture. A mixture container must have two valves, liquid and gas.
- Evacuate the empty container using a vacuum pump to remove any remaining gas and moisture in the container.

- Calculate the total mass of the mixture to blend and its component mass. For example, use 5 kg of R32 and 5 kg of R125 to make 10 kg of R32/R125 mixture with 50/50 wt.% mass fraction.
- Charge lower vapor pressure refrigerants first. If the higher vapor pressure refrigerant is charged first, it may not be easy to charge the lower vapor pressure fluid to the mixture container.
- To enhance the refrigerant transfer, place the container for the mixture in a cold place (such as in a bucket filled with dry ice or ice water).
- Place a R125 container on the weighing scale and connect the R125 container and the mixture container with a refrigerant hose.
- While weighing the R125 container, transfer R125 to the mixture container by adjusting container valves until the required amount of R125 is transferred.
- Repeat the same procedure to transfer R32 from the R32 container to the mixture container.
- Mix well by shaking.
- Take a liquid sample of the refrigerant mixture from the mixture container and check the mass fraction of the mixture by using a gas chromatograph.
- Adjust the mass fraction. If the desired mass fraction is not achieved, then calculate the amount to add from the measured mass fraction, add the insufficient component and measure the mass fraction again. Repeat this procedure until the desired mass fraction is achieved within 1%.

9.5 REFRIGERANT MASS FRACTION MEASUREMENT

Maintaining the refrigerant mass fraction for mixtures is important to guarantee design performance, safety and reliability. The procedure used for measuring the refrigerant mass fraction follows.

- Prepare refrigerant sample bottle. The volume of the sample bottle is sized such that after the refrigerant sample expands into the bottle, only vapor exists (approximately 1 l bottle is used as an example).
- Clean the sample bottles using acetone or another solvent to remove any residual refrigerant and oil that might accumulate in the bottles: Evacuate the bottles and pour approximately 100 ml of acetone in the bottle at atmospheric pressure. Violently agitate the sample bottle for approximately 30 seconds. Charge nitrogen into the bottle to about 350 kPa (nitrogen is used to rapidly push the acetone–oil solution out of the sample bottle). Turn the bottle upside down and open the valve to allow the acetone–oil solution to flow out.
- Evacuate the (preferably heated) sample bottle for several hours to remove any remaining acetone.
- Attach the sample bottle to the sample tube already installed to a system tube. The sample tube must be inserted into the middle of the refrigerant stream in the liquid and vapor lines to ensure that bulk samples are

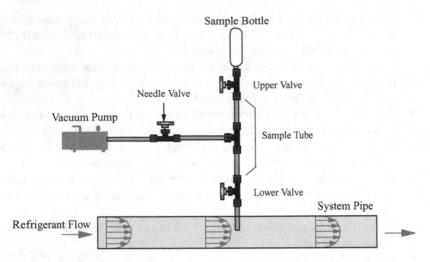

FIGURE 9.5.1 Sampling refrigerant.

analyzed as shown in Figure 9.5.1. To control the sampling, it is common practice to install either a manual valve or a solenoid valve between the sample bottle and tube.

- Take the first refrigerant sample into the sample bottle by opening valves and throw away to flush the sample section.
- Reconnect a new sample bottle and evacuate the sample section by using the vacuum pump.
- Take the refrigerant sample into the sample tube by opening and closing the lower valve. To ensure that the sample tube is filled with liquid, it should be packed in dry ice or ice water.
- Expand the refrigerant sample from the tube into the sample bottle by opening and closing the upper valve. The sample tube should be heated to ensure that all refrigerant is completely evaporated.
- Remove the sample bottle from the system after closing the valves.
- Analyze the samples by using the gas chromatograph.
- Repeat the above procedure three times to ensure that the device is working properly.

9.6   EVACUATING A SYSTEM

Evacuating refrigeration and air-conditioning systems is very important to secure a steady performance and reliability of the system since this process removes residual noncondensable gases and moisture from the system. Evacuating can remove gases quickly but takes longer to remove moisture from the system. POE oils are very hydroscopic and form acids together with the moisture and refrigerant at elevated temperature. General procedures are as follows:

FIGURE 9.6.1 Vacuum pump. (Photo courtesy of INFICON.)

- Choose the size of vacuum pump that is appropriate for the system size. Make sure the system pressure is not higher than 1 atm and that all refrigerant that may have been in the system is reclaimed. Figure 9.6.1 shows a typical 1/2 hp deep vacuum pump.
- Make sure the space around the pump exhaust is well ventilated since the vacuum pump will emit residual gases from the system and sometimes discharge oil mist.
- Connect the middle port of the manifold gauge set to the vacuum pump.
- Open both system's service ports (liquid and gas line ports).
- Open both low and high pressure side valves of the manifold gauge set.
- Turn on the vacuum pump. Keep evacuating until the vacuum pressure reaches 500 μm of Hg, then run the vacuum pump at least 30 minutes more to remove moisture left in the system or dissolved in the oils. The longer the time for evacuation and the higher the temperature of all internal surfaces of the system, the more complete is the removal of moisture.
- Shut down procedure for vacuum: Close valve to the system to be evacuated, bring the inlet port pump to atmospheric pressure (open valve or disconnect hose), then pump can be turned off.
- Vacuum pump oil change: After every 20 to 30 operation hours, the oil of the vacuum pump must be changed as it picks up moisture and contamination. Keep a service sheet for every vacuum pump where operating hours and oil changes are recorded.

9.7 REFRIGERANT CHARGE

Refrigerant containers are different for pure and mixed refrigerants to ensure that mixtures maintain consistent mass fractions. Containers for pure refrigerants such as R22 and R134a have only one discharge port since charging liquid or gas does

FIGURE 9.7.1 Manifold gauge set. (Photo courtesy of INFICON.)

not make any difference. However, containers for mixtures such as R407C and R410A have two discharge ports, a liquid port and a gas port. When charging refrigerant mixtures to the system, the liquid port should be used to maintain the mass fraction. Other than this precaution, the same refrigerant charging procedure can be used for both pure refrigerants and mixtures, explained as follows:

- Determine the required refrigerant charge from the product specification label or manual. If not found, contact the system manufacturer. If all else fails, the degre of superheating at the evaporater outlet (or possibly subcooling at the condensor outlet) can be used as a charging criterion. However, this method is not ideal since the charging progress does not usually occur at design conditions.
- Place a refrigerant container on the weighing scale and connect the refrigerant container, liquid port, system service ports and manifold gauge set (Figure 9.7.1) with refrigerant hoses.
- Make sure the system has been evacuated. If not, evacuate the system based on Section 9.6.
- Purge the hose and manifold gauge set if necessary (emission minimized hoses do not need this procedure). Add a small amount of refrigerant gas by quickly opening and closing a refrigerant container's gas-side valve. Release a refrigerant and air mixture in the hose by loosening and tightening the refrigerant hose.
- Open a system liquid line service port.
- Start to charge liquid refrigerant to the system by opening the refrigerant container's liquid-side valve very slightly while the system stays off.

- Add approximately 70 to 80% of the required charge to the system in liquid form by using the pressure difference between the refrigerant container and the system.
- Switch the manifold gauge valve opening. Close the high pressure side valve and open the low pressure side valve. Open a system gas line service port. Add more refrigerant to obtain the design charge by operating the compressor.
- If the optimal charge is uncertain, add more refrigerant until the sight glass at the condenser outlet becomes clear, securing a degree of subcooling at the condenser outlet or the desired superheating at the evaporator outlet is obtained.

9.8 LEAK CHECKING

There are three common ways to check a system for leaks: bubble test with soap solution, electronic leak detector and system pressure change observation.

- Charge the system with nitrogen up to the system design pressure.
- Bubble test:
  - Prepare leak detection soap solution for leak testing.
  - Apply soap solution to all fittings and watch carefully for bubbles. Leak location is indicated by soap bubbles.
  - This method is a useful way to find a relatively large leak.

- Electronic leak detector (Figure 9.8.1):
  - Add 50 to 100 g HCFC or HFC refrigerants to the system. If available, use the recovered refrigerant. Pressurize with nitrogen.
  - Move the probe slowly around all fittings. Locate the leak point when the leak detector signals a leak.
  - This method is a useful way to find a relatively small leak.

- Pressure change observation:
  - Charge nitrogen up to the desired test pressure.
  - Measure the system pressure over the desired time (usually overnight). Pressure should remain unchanged. Temperature changes of the surroundings must be considered.
  - This method is a useful way to find a relatively small leak.

9.9 MASS FRACTION SHIFTS

An inherent feature of zeotropic mixtures is the different boiling points of their components that result in different vapor and liquid mass fractions at equilibrium. As shown in Figure 9.9.1, the saturated vapor contains more low boiling point component while the saturated liquid contains more high boiling point component

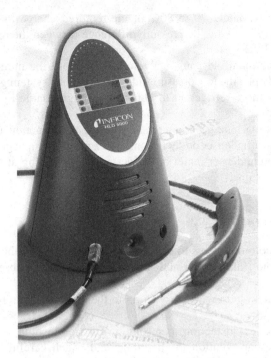

FIGURE 9.8.1 Electronic leak detector. (Photo courtesy of INFICON.)

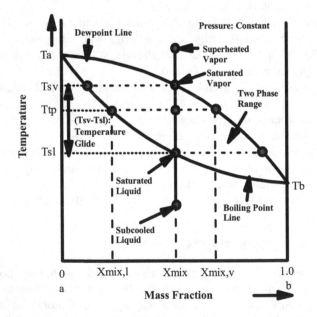

FIGURE 9.9.1 Temperature–mass fraction diagram of a zeotropic mixture.

when the temperature is between its saturated liquid and vapor temperature (T_{sl} and T_{sv}). This mass fraction difference brings new design issues:

- Mass fraction shift within the cycle.
- Mass fraction shift due to leakage.
- Mass fraction shift during refrigerant transfer between containers.
- Cycle performance effects due to the mass fraction shift.

9.9.1 MASS FRACTION SHIFT IN THE CYCLE

Two major mechanisms cause a mass fraction shift within the cycle: the use of an accumulator due to the vapor-liquid volume fractionation effect and the differential refrigerant solubility into the compressor lubricant. Three different investigation efforts are introduced in this section.

To investigate this issue, Corr and Murphy (1994) developed a simplified bulk-tank model, "Cyclops." Figure 9.9.2 shows the refrigerant distribution in the cycle. The compressor contains oil and refrigerant mixture in the oil sump and refrigerant gas within the shell other than the sump. The condenser contains two-phase, super-heated vapor and subcooled liquid refrigerant. The evaporator contains two-phase and superheated refrigerant.

Usually the internal volume of the expansion device is negligible. When the cycle reaches its steady state, the circulating mass fraction should be the same at every inlet and outlet of each component. When evaporating and condensing processes are assumed as isobaric processes, the mass fraction of the two-phase refrigerant differs from that of the single phase refrigerants (either superheated gas or subcooled liquid) due to the different boiling or dew points of the pure components. If the evaporation and condensation processes are simplified as bulk-liquid

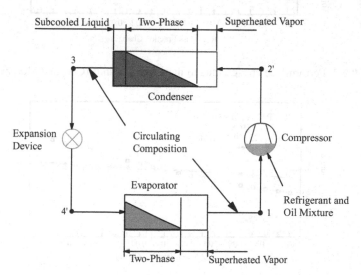

FIGURE 9.9.2 Refrigerant distribution in the cycle.

and bulk-vapor, the average liquid mass fraction and the average vapor mass fraction are different from the circulating mass fraction due to the vapor-liquid volume fractionation. Moreover, the circulating mass fraction shifts from the charged mass fraction because the more volatile component accumulates predominantly in the condenser vapor phase.

Corr and Murphy (1994) assumed that the subcooled liquid mass fraction is similar to the average saturated liquid mass fraction of the condenser and the superheated suction vapor mass fraction is similar to the average saturated vapor mass fraction of the evaporator. With this simplified assumption, Corr and Murphy showed that −1 to 3% of R32 mass fraction is shifted in the circulating refrigerant when R32/R125 (60/40 wt.%) is charged.

To illustrate the effect of mass fraction shift on the cycle performance, the changes of volumetric capacity and coefficient of performance (COP) are calculated when R32 mass fractions of two binary mixtures of R32/R134a (30/70 wt.%) and R32/R125 (50/50 wt.%) are changed with 10% variation from the charged concentration, as shown in Figures 9.9.3 and 9.9.4. R32/R134a (30/70 wt.%) exhibits approximately twice larger changes in volumetric capacity and COP due to the mass fraction shift than that of R32/R125 (50/50 wt.%). If the circulating mass fraction

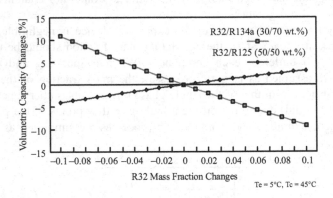

FIGURE 9.9.3 Performance changes due to mass fraction shift (Corr and Murphy, 1994).

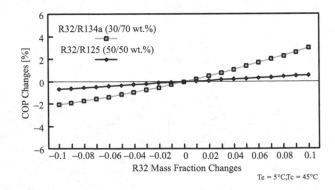

FIGURE 9.9.4 Performance changes due to mass fraction shift (Corr and Murphy, 1994).

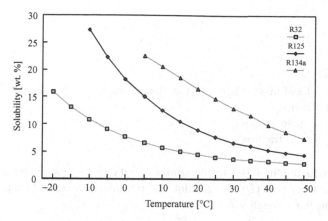

FIGURE 9.9.5 Solubility of HFC refrigerants in ISO 32 polyol-ester (Corr and Murphy, 1994).

shift ranges 10%, R32/R134a (30/70 wt.%) undergoes volumetric capacity and COP variation within 10% and 2%, respectively. This result is due to the larger temperature glide of R32/R134a (30/70 wt.%), a zeotropic mixture, than that of R32/R125 (50/50 wt.%), almost an azeotropic mixture. Therefore, the mass fraction shift in the cycle should be accounted for in the cycle design, especially for the zeotropic mixtures with large temperature glides.

Figure 9.9.5 shows an example for the different HFC refrigerant solubility in the ISO 32 polyolester oil (Corr and Murphy, 1994). As shown in Figure 9.9.5, the solubility of R32 is much lower than that of R125 and R134a due to its low molecular weight. Therefore, R32 resides in the heat exchangers or circulates through the cycle, more so than the other components do if binary mixtures, R32/R125 or R32/R134a, are charged. As the compressor sump temperature decreases, the difference between the refrigerant solubility increases. Even more R32 then migrates to the heat exchangers or circulates in the cycle. However, the impact of the different refrigerant solubility in the oil is not so significant because for large systems, the oil charge is usually smaller than that of the refrigerant charge and the compressor sump temperature is usually high enough to maintain a lower refrigerant solubility. In small systems such as domestic refrigerators, the oil charge can equal or exceed the refrigerant charge and these effects are more important.

In 1995, Sumida et al. also investigated the circulating mass fraction shift of a zeotropic mixture in the cycle. Sumida et al. distinguished between "circulating mass fraction" on a mass flux basis from "existing mass fraction" on a mass basis, defined as: The *circulating mass fraction* (α) is the ratio of the mass flux of the more volatile component to the total mass flux of the refrigerant mixture in the cycle, as shown in Equation 9.1. The *existing mass fraction* (β) is the ratio of the mass of the more volatile component to the total mass of the refrigerant mixture in the cycle, as shown in Equation 9.2.

$$\alpha = \frac{w_l(1-f_v)\rho_l u_l + w_v f_v \rho_v u_v}{(1-f_v)\rho_l u_l + f_v \rho_v u_v} \tag{9.1}$$

$$\beta = \frac{w_l(1-f_v)\rho_l + w_v f_v \rho_v}{(1-f_v)\rho_l + f_v \rho_v} \tag{9.2}$$

where

w = local mass fraction of the more volatile component
ρ = density
u = mean velocity
f_g = void fraction

By introducing the following nondimensional ratios — gas-liquid area ratio (F), gas-liquid density ratio (D), and gas-liquid slip ratio (S) — α and β can be expressed as Equations 9.3 through 9.7.

$$F = f_v/(1-f_v) \tag{9.3}$$

$$D = \rho_v/\rho_l \tag{9.4}$$

$$S = u_v/u_l \tag{9.5}$$

$$\alpha = \frac{w_l + w_v FDS}{1 + FDS} \tag{9.6}$$

$$\beta = \frac{w_l + w_v FD}{1 + FD} \tag{9.7}$$

If δ is defined as the mass fraction difference between the circulating mass fraction and existing mass fraction, δ can be expressed as Equation 9.8,

$$\delta = \alpha - \beta = \frac{FD(S-1)(w_g - w_l)}{(1+FDS)(1+FD)} \tag{9.8}$$

Therefore, the circulating mass fraction is different from the existing mass fraction except in the case of either a zero slip between gas and liquid or the same liquid and vapor mass fraction. Since two-phase flow of any fluid has a positive slip ratio and since zeotropic mixtures have a higher gas mass fraction than liquid mass fraction, the mass fraction difference δ always becomes positive for the two-phase flow of a zeotropic mixture.

Consider the mass fraction difference between the charged mass fraction (γ) and the circulating mass fraction (α). From the mass conservation of the more volatile component in the total cycle, γ can be expressed in terms of β as shown in Equation 9.9,

$$\gamma W_{total} = \sum (\beta_i W_i) \tag{9.9}$$

If ϕ is defined as the mass fraction difference between the circulating mass fraction and charged mass fraction, ϕ can be expressed in terms of δ as shown in Equation 9.10 by rearranging Equation 9.9 with Equation 9.8,

$$\phi = \alpha - \gamma = \sum_i \left(\delta_i \frac{W_i}{W_{total}} \right) \tag{9.10}$$

To determine the mass fraction difference between the circulating mass fraction and charged mass fraction ϕ, the mass fraction difference between the circulating mass fraction and existing mass fraction δ must be determined for each component. However, δ for single phase becomes zero because of zero slip; therefore, only the contribution of the two-phase components remains. From the void fraction correlation for f_g, parameters F and S are determined from the Equations 9.3 and 9.11,

$$S = \frac{1 - f_v}{f_v (1/x - 1)(\rho_v/\rho_l)} \tag{9.11}$$

Sumida et al. used Smith's correlation for void fraction and showed that the circulating mass fraction of R32/134a (30/70 wt.%) shifts from 0.32 to 0.42, depending on the amount of refrigerant in the accumulator. They also showed that the capacity is affected up to 10% because of this mass fraction shift; however, the COP is maintained.

In another study of this issue, Chen and Kruse (1995) also investigated the circulation mass fraction shift. Chen and Kruse recommended Highmark's void fraction model after comparing seven different void fraction models with their experiment. Chen and Kruse examined the mass fraction shift of R23/R152a (20/80 and 25/75 wt.%) and found the mass fraction shifts of 0.06 and 0.05 for each mass fraction. Chen and Kruse also showed the mass fraction shifts and their performance effects for three additional refrigerant mixtures, as shown in Table 9.1.

TABLE 9.1
Mass Fraction Shift and Capacity Changes of Three Refrigerant Mixtures

Refrigerant Mixture	Mass Fraction Shift [wt.%]	Capacity Changes
R407C (R32/R125/R134a, 23/25/52 wt.%)	R32: +1.5, R125: +1.4, R134a: –3.2	+0.9%
R32/R134a (30/70 wt.%)	R32: +3.0, R134a: –3.0	+1.1%
R404A (R125/R143a, 44/52 wt.%)	R125: +0.1, R143a: +0.1, R134a: –0.2	+0.03%

Source: Chen and Kruse, 1995.

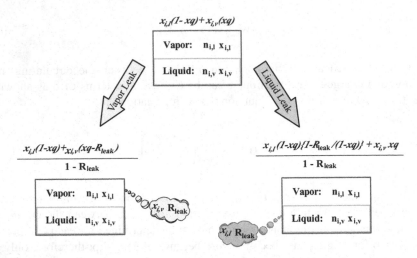

FIGURE 9.9.6 Refrigerant leakage model (Kim and Didion, 1995a).

9.9.2 MASS FRACTION SHIFT DUE TO LEAKAGE

The simplest and biggest problem of the refrigeration and air-conditioning equipment is a refrigerant leak from refrigerant tanks or refrigeration systems. In the past, the system has been simply charged again or refilled on site by adjusting charge to obtain the desired operating pressures in the presence of a leak. However, this method of recharging needs to be reexamined especially when the system uses refrigerant mixtures because of the mass fraction shift. Refrigerant leaks can be divided into a vapor leak and liquid leak, as shown in Figure 9.9.6. Typically, the vapor leak is a slow process of losing refrigerant since the vapor density is much lower than that of liquid. Therefore, the vapor leak can be assumed as an isothermal leak process since the refrigerant remaining in the refrigerant tank or refrigeration system can exchange heat with the ambient and maintain a constant temperature. The liquid leak is a fast process of losing refrigerant and can be assumed as an adiabatic leak process since the temperature of the refrigerant remaining in the refrigerant tank or refrigeration system can drop quickly without exchanging heat with the ambient.

Isothermal vapor leaks and adiabatic liquid leaks can be modeled based on the following assumptions:

- Refrigerant mixture is in a thermodynamic equilibrium condition.
- Only one phase of refrigerants leaks.
- Mass fraction of leaking refrigerant is the same as that of the phase that is leaking.

Liquid and vapor are assumed initially in thermodynamic equilibrium. If x_l, x_v and x_q are defined as the mass faction of the liquid phase, mass faction of the vapor phase and overall vapor mass quality, the overall mass fraction of the i-th component in the tank or system can be expressed as Equation 9.12,

$$x_i = x_{l,i}(1 - x_q) + x_{v,i}x_q \tag{9.12}$$

Consider the vapor leak by introducing a nondimensional parameter, leakage rate (R_{leak}), defined as the mass fraction of leaking vapor to the total mass. The new overall mass fraction of the i-th component in the tank or system can then be expressed as Equation 9.13,

$$x_{i,new} = \frac{x_{l,i}(1 - x_q) + x_{v,i}x_q(1 - R_{leak}/x_q)}{1 - R_{leak}} \tag{9.13}$$

Therefore, the change of the overall mass fraction of the i-th component in the tank or system can be expressed as Equation 9.14,

$$\Delta x_i = x_{i,new} - x_i = \frac{R_{leak}}{1 - R_{leak}}(1 - x_q)(x_{l,i} - x_{v,i}) \tag{9.14}$$

Similarly, if there is a liquid leak, the change of the overall mass fraction of the i-th component in the tank or system can be expressed as Equation 9.15,

$$\Delta x_i = x_{i,new} - x_i = \frac{R_{leak}}{1 - R_{leak}}x_q(x_{v,i} - x_{l,i}) \tag{9.15}$$

The isothermal vapor leak is of more concern than the liquid leak because it occurs for a longer period of time and sometimes is harder to detect; therefore, a partial recharging may be considered after the vapor leak. If there is a liquid leak, it occurs over a short period of time and the loss of total refrigerant is significant; therefore, a complete recharging may be considered after a liquid leak. However, the isothermal vapor leak also needs to be treated carefully when the components of the mixture have a large vapor pressure difference because the mass fraction shift becomes the issue in this case.

Kim and Didion (1995a) developed a leakage simulation model, REFLEAK, using the approach described above. Figure 9.9.7 shows the simulation results of an isothermal vapor leak cases for the selected refrigerants R410A, R32/R125 (30/70 wt.%) and R407C by using REFLEAK. The boiling points of R32, R125, and R134a are −51.7, −48.1, and −26.1°C, respectively. As shown in Figure 9.9.7, the mass fraction of the more volatile component decreases and the mass fraction of the less volatile component increases as a result of a vapor leak. If there is 50% vapor lost, R410A loses 2% of R32 liquid and 1% of R32 vapor. For the same case, R32/R125 (30/70 wt.%) loses 2% of R32 liquid and vapor and R407C loses 7% of R32 liquid and vapor. Therefore, the impact of the vapor leak will be more severe for R407C and the next one is R32/R125 (30/70 wt.%). R410A has the least impact.

Bivens et al. (1996) also developed a similar model by integrating the differential equations for mass fraction shifts and obtained a similar result as Kim and Didion.

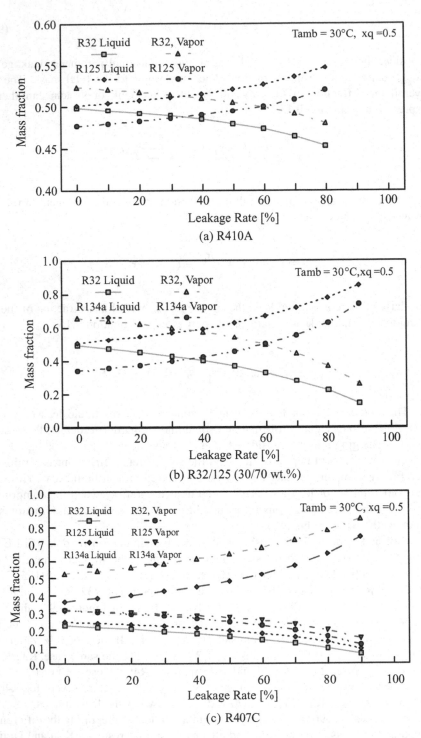

FIGURE 9.9.7 Isothermal vapor leak process (Kim and Didion, 1995b).

TABLE 9.2
Mass Fraction Shift during Liquid Charging from Container (25°C)

Liquid Level [%]	R410A		R407C		
	R32	R125	R32	R125	R134a
85	49.50	50.50	23.0	25.0	52.0
50	49.42	50.58	22.8	24.9	52.3
40	49.39	50.61	22.7	24.8	52.5
30	49.35	50.65	22.6	24.7	52.7
20	49.30	50.70	22.5	24.6	52.9
15	49.26	50.74	22.3	24.6	53.1
10	49.20	50.80	22.2	24.5	53.3
5	49.15	50.85	22.0	24.4	53.6
2	49.10	50.90	21.9	24.2	53.9

Source: Bivens and Yokozeki, 1998.

9.9.3 MASS FRACTION SHIFT DURING CONTAINER TRANSFERS

In general, refrigerants undergo a couple of transportation steps between containers before they are charged to the final system. While being transferred, there is a potential mass fraction shift. When charging refrigerant mixtures, it is important to charge from the liquid port of the storage container to minimize any mass fraction shift. Therefore, emptying and refilling a container can be regarded as the same process as that of an adiabatic liquid leak. Bivens and Yokozeki (1998) examined this issue in detail by using the model they developed previously.

Table 9.2 shows mass fraction shifts of two mixtures while the container is being emptied from 100% liquid to 2% liquid. As can be seen from Table 9.2, the maximum charging mass fraction shift (the mass fraction of the mixture that is actually charged to a refrigeration system from the container) is 0.4% of R32 and R125 for R410A and 1.9% of R134a for R407C.

REFERENCES

Air-Conditioning and Refrigeration Institute, 1994, *Applicant Study Guide for Refrigerant Containment*, Section 608, Clean Air Act, 2nd Edition.

Bivens, D.B., M.B. Shiflett, C.C. Allgood and A. Yokozeki, 1996, "Zeotropic Mixture Separations Analyses," Proceedings of the 1996 International Refrigeration Conference at Purdue, West Lafayette, IN, pp. 113–118.

Bivens, D.B. and A. Yokozeki, 1998, "Mass Fraction Changes during Container Transfers for Multi-component Refrigerants," Proceedings of the 1998 International Refrigeration Conference at Purdue, West Lafayette, IN, pp. 81–86.

Chen, J. and H. Kruse, 1995, "Calculating Circulation Concentration of Zeotropic Refrigerant Mixtures," *International Journal of HVAC&R Research*, Vol. 1, No. 3, pp. 219–231.

Chen, J. and H. Kruse, 1997, "Concentration Shift Simulation for the Mixed Refrigerants R404A, R32/134a, and R407C in an Air-Conditioning System," *International Journal of HVAC&R Research*, Vol. 3, No. 2, pp. 149–157.

Corr, S. and F.T. Murphy, 1994, "Mass Fraction Shifts of Zeotropic HFC Refrigerants in Service," Proceedings of IIR Conference for CFCs, The Day After, pp. 29–40.

Kim, M.S. and D.A. Didion, 1995a, "Simulation of Isothermal and Adiabatic Leak Processes of Zeotropic Refrigerant Mixtures," *International Journal of HVAC&R Research*, Vol. 1, No. 1, pp. 3–20.

Kim, M.S. and D.A. Didion, 1995b, "Simulation of a Leak/Recharge Process of Refrigerant Mixtures," *International Journal of HVAC&R Research*, Vol. 1, No. 3, pp. 242–254.

Sumida, Y., N. Tanaka and T. Okazaki, 1995, "Prediction of the Circulating Mass fraction of a Zeotropic Mixture in a Refrigerant Cycle," Proceedings of 19th International Congress of Refrigeration, Vol. IVb, pp. 1013–1020.

Index

Printed in the United States
by Baker & Taylor Publisher Services